NASA SP-214

APOLLO 11

Preliminary Science Report

Scientific and Technical Information Division
OFFICE OF TECHNOLOGY UTILIZATION   1969
NATIONAL AERONAUTICS AND SPACE ADMINISTRATION
Washington, D.C.

ACKNOWLEDGMENT

The material submitted for the Lunar Surface Scientific Mission of Apollo 11 was reviewed by a NASA Manned Spacecraft Center Technical Review Board consisting of the following members: J. M. West (Chairman), P. R. Bell, A. J. Calio, J. W. Harris, H. H. Schmitt, S. H. Simpkinson, W. K. Stephenson, and D. G. Wiseman.

Foreword

OUR FIRST JOURNEY to the Moon ushered in a new era in which man will no longer be confined to his home planet. The concept of traveling across the vastness of space to new worlds has stirred the imagination of men everywhere. One-sixth of the Earth's population watched as the Apollo 11 astronauts walked and worked a quarter of a million miles away.

The success of this mission has opened new fields of exploration and research — research which will lead to a greater understanding of our planet and provide a new insight into the origin and history of the solar system. The Apollo 11 mission was only a beginning, however. Subsequent missions will reflect more ambitious scientific objectives and will include more comprehensive observations and measurements at a variety of lunar sites.

This document is a preliminary report of the initial scientific observations resulting from the Apollo 11 mission. We expect that further significant results will come from more detailed analysis of the returned samples of lunar material, and from additional study of the photographs and data obtained from the emplaced experiments. Beyond that, we look forward to increasing international participation in the exploration of the Moon and neighboring regions of our solar system.

THOMAS O. PAINE
Administrator
National Aeronautics and Space Administration

OCTOBER 31, 1969

Contents

		PAGE
	INTRODUCTION Robert R. Gilruth and George M. Low	vii
	SUMMARY OF SCIENTIFIC RESULTS W. N. Hess and A. J. Calio	1
1	PHOTOGRAPHIC SUMMARY OF APOLLO 11 MISSION James H. Sasser	9
2	CREW OBSERVATIONS Edwin E. Aldrin, Jr., Neil A. Armstrong, and Michael Collins	35
3	GEOLOGIC SETTING OF THE LUNAR SAMPLES RETURNED BY THE APOLLO 11 MISSION E. M. Shoemaker, N. G. Bailey, R. M. Batson, D. H. Dahlem, T. H. Foss, M. J. Grolier, E. N. Goddard, M. H. Hait, H. E. Holt, K. B. Larson, J. J. Rennilson, G. G. Schaber, D. L. Schleicher, H. H. Schmitt, R. L. Sutton, G. A. Swann, A. C. Waters, and M. N. West	41
4	APOLLO 11 SOIL MECHANICS INVESTIGATION N. C. Costes, W. D. Carrier, J. K. Mitchell, and R. F. Scott	85
5	PRELIMINARY EXAMINATION OF LUNAR SAMPLES	123
6	PASSIVE SEISMIC EXPERIMENT Gary V. Latham, Maurice Ewing, Frank Press, George Sutton, James Dorman, Nafi Toksoz, Ralph Wiggins, Yosio Nakamura, John Derr, and Frederick Duennebier	143
7	LASER RANGING RETROREFLECTOR C. O. Alley, P. L. Bender, R. F. Chang, D. G. Currie, R. H. Dicke, J. E. Faller, W. M. Kaula, G. J. F. MacDonald, J. D. Mulholland, H. H. Plotkin, S. K. Poultney, D. T. Wilkinson, Irvin Winer, Walter Carrion, Tom Johnson, Paul Spadin, Lloyd Robinson, E. Joseph Wampler, Donald Wieber, E. Silverberg, C. Steggerda, J. Mullendore, J. Rayner, W. Williams, Brian Warner, Harvey Richardson, and B. Bopp	163
8	THE SOLAR-WIND COMPOSITION EXPERIMENT J. Geiss, P. Eberhardt, P. Signer, F. Buehler, and J. Meister	183
9	LUNAR SURFACE CLOSEUP STEREOSCOPIC PHOTOGRAPHY	187
10	THE MODIFIED DUST DETECTOR IN THE EARLY APOLLO SCIENTIFIC EXPERIMENTS PACKAGE J. R. Bates, S. C. Freden, and B. J. O'Brien	199
	APPENDIX A — GLOSSARY OF TERMS	203
	APPENDIX B — ACRONYMS	204

Introduction

Robert R. Gilruth and George M. Low

THE MANNED SPACE FLIGHT PROGRAM is dedicated to the exploration and use of space by man. In the immediate future, men will continue exploring the Moon, adding greatly to our knowledge of the Moon, the Earth, and the solar system. It is expected that in Earth orbital laboratories men will conduct experiments and make observations that are possible only in the space environment.

In standing firmly behind the space science program, it has been necessary to provide means for man to live and work in "this new ocean." A transportation system has had to be provided with a navigation system of great precision. An extensive medical program has been necessary to provide data on man's reaction to the space environment. The Mercury, Gemini, and early Apollo missions have produced, step by step, the answers needed for the Apollo 11 lunar landing mission.

The rapid progress of the program obscured its great problems. The success of the Apollo 11 mission was solidly based on excellent technology, sound decisions, and a test program that was carefully planned and executed. To this foundation was added the skill and bravery of the astronauts, backed up by a fully trained and highly motivated ground team. It must not be forgotten that the lunar mission was very complex from all points of view—planning, hardware, software, and operations. Of necessity, margins were small and even small deviations in performance or conduct of the mission could have jeopardized mission success.

In parallel with the emphasis on engineering problems and their solution, the scientific part of the Apollo 11 mission was planned and executed with great care. The samples of lunar soil and rocks returned by the astronauts will add much detailed scientific information. The photographs and observations of the crew have already answered some questions man has asked for thousands of years. The emplaced experiments have yielded data unavailable until now. This report is preliminary and covers only the initial scientific results of the Apollo 11 mission. Much work remains for the large number of scientists involved to understand and interpret the facts that are only partly exposed today.

Summary of Scientific Results

W. N. Hess and A. J. Calio

The scientific objectives of the Apollo 11 mission, in order of priority, were the following:

(1) To collect early in the extravehicular activity (EVA) a sample, called the contingency sample, of approximately 1 kg of lunar surface material to insure that some lunar material would be returned to Earth.

(2) To fill rapidly one of the two sample return containers with approximately 10 kg of the lunar material, called the bulk sample, to insure the return of an adequate amount of material to meet the needs of the principal investigators.

(3) To deploy three experiments on the lunar surface:

    (a) A passive seismometer to study lunar seismic events, the Passive Seismic Experiment Package (PSEP)

    (b) An optical corner reflector to study lunar librations, the Laser Ranging Retroreflector (LRRR)

    (c) A solar-wind composition (SWC) experiment to measure the types and energies of the solar wind on the lunar surface

4) To fill the second sample return container with carefully selected lunar material placed into the local geologic context, to drive two core tubes into the surface, and to return the tubes with the stratigraphically organized material, called the documented sample.

During these tasks, photographs of the surface were to be taken using a 70-mm Hasselblad camera and a closeup stereoscopic camera (the Apollo Lunar Surface Closeup Camera (ALSCC)). The scientific tasks and a variety of other tasks were planned for a 2-hr and 40-min time period.

According to mission plans, the time allotted for the collection of the documented sample, which had the lowest priority, would be shortened if time were insufficient. All the scientific tasks were completed satisfactorily, all instruments were deployed, and approximately 20 kg of lunar material were returned to Earth. The documented-sample period was extremely short, however, and samples collected during this period were not carefully photographed in place or documented in other ways.

Nature of the Lunar Surface

The Apollo 11 lunar module (LM) landed in the southwestern part of Mare Tranquillitatis, approximately 50 km from the closest highland material and approximately 400 m west of a sharp-rimmed blocky crater approximately 180 m in diameter. Rays of ejecta from this crater extend past the landing site. Rays from more distant craters, including the crater Theophilus, are also in the landing region.

Surface material at the landing site consists of unsorted fragmental debris ranging in size from approximately 1 m to microscopic particles, which make up the majority of the material. This debris layer, the regolith, is approximately 5 m thick in the region near the landing site, as judged by the blockiness of material near various-sized craters.

The soil on the lunar surface is weakly cohesive, as shown by the ability of the soil to stand on vertical slopes. The fine grains tend to stick together, precluding clods of material that crumpled under the astronauts' boots. The depth of the astronauts' footprints and the penetration of the LM landing gear correspond to static bearing pressures of approximately 1 psi. The surfaces were relatively soft to depths of 5 to 20 cm. Deeper than these depths, the resistance of the material to penetration increases considerably. In general, the lunar soil at the landing site was similar in appearance, behavior, and mechanical properties to the soil encountered at the Surveyor equatorial landing sites. Although the lunar soil differs considerably in composition and in range of particle shapes from a terrestrial soil of the same particle-size distribution, the lunar soil does not appear to differ significantly from similar terrestrial soil in mechanical behavior.

Both rounded and angular rocks appear on the surface in profusion. All degrees of burial are present, and fillets on the sides of rocks, caused by the powdery surface material being piled up by some erosional process, were common. At least three of the rocks returned to Earth have been identified in photographs of the lunar surface. On what appears to be the upper surface of several rocks, a thin rind of altered material approximately 1 mm thick is found. This rind is lighter colored than the remainder of the rock and appears to be caused by shattering of mineral grains.

One outstanding feature of the surfaces of the rocks returned to Earth is the existence of several glass-lined pits 1 mm or smaller in diameter. These glass-lined pits appear only on the surfaces of rocks. (More of these pits are observed on the top surfaces of rocks with known orientations.) Quite clearly, these pits are of external origin. The glass overlaps of the surface of the rock and the resulting features clearly resemble hypervelocity impact craters. However, the pits do not resemble craters made in the laboratory by hypervelocity particles, and the origin of the pits is presently unknown.

The most interesting and unexpected surface features discovered and photographed by the astronauts are glassy patches on the lunar surface that are described by the astronauts as resembling drops of solder. These patches were observed only inside several raised-rim craters approximately 1 m in diameter. These glassy blebs may be formed by low-velocity molten material splattering into the craters, or they may be formed from material that has been melted in place. The section of this document entitled "Lunar Surface Closeup Stereoscopic Photography" presents an interesting theory of the origin of the blebs, based on radiation heating. This theory postulates that within the last 100 000 years, the Sun had a superflare or mininova event that heated the lunar surface to a temperature that caused material inside the craters to melt, but did not cause surface material to melt. According to the theory, the reason that material inside the craters melted while surface material did not is that a focusing effect caused the temperature inside the craters to increase. This radiation-heating theory is certainly not yet proved, but no other plausible theories have been advanced to explain the blebs' being located only in the bottoms of craters. None of the blebs photographed by the astronauts were returned in the sample containers, and no blebs have been identified in the samples.

The Passive Seismic Experiment

Since the time of the Ranger 1 mission, scientists have been trying to land a seismometer on the surface of the Moon to search for moonquakes. Successful operation of a seismometer is extremely important for understanding of the internal structure of a planet and to a search for possible layering or discontinuities. On the Apollo 11 mission, a seismometer was placed on the surface of the Moon, and the instrument operated satisfactorily for 21 days. The instrument contained four separate components. Three long-period (LP) (approximately 15-sec resonance) seismometers were alined orthogonally to measure surface motion both horizontally and vertically. A single-axis, short-period (SP) seismometer, with a resonant period of approximately 1 sec, measured vertical motion. The system had tilt adjustment motors to level the system upon command from Earth. The instrument was deployed on the lunar surface approximately 16 m from the LM and was turned on while the astronauts were on the lunar surface. Signals were received when the crewmen climbed the LM ladder, used a hammer to pound on the core tubes, and jettisoned equipment, including the portable life support systems (PLSS).

Actual maximum instrument temperature (approximately 190° F) exceeded the planned maximum instrument temperature by approximately 50° F. Even at this elevated temperature, the instrument worked satisfactorily during the first lunar day and during part of the second. However, near noon of the second lunar day, the instrument no longer accepted commands from Earth stations; therefore, the experiment was terminated.

The LM appears to have been a rich source of seismic noise. One class of repeating signals of nearly identical structure that was observed gradually died out over a period of many Earth days. These signals seem to be related to events on the LM such as fuel venting, valve chatter, or other mechanical motion. The signals had a dominant frequency of 7.2 Hz before LM ascent and of 8.0 Hz following ascent. These frequencies appear to be characteristic frequencies of the LM structure.

Several events showing dispersion and having the appearance of surface waves were detected on the LP seismometers. These wave trains often occurred simultaneously with a series of pulses on the SP seismometer. It is not yet certain, but these waves are probably not the result of real seismic events but are of instrumental origin.

Several other classes of events of unknown origins were observed. The following are possible source mechanisms for these observed seismic signals:

(1) Venting gases from the LM and circulating fluids within the LM

(2) Thermoelastic stress relief within the LM and the PSEP

(3) Meteoroid impacts on the LM, the PSEP, and the lunar surface

(4) Displacement of rock material along steep crater slopes

(5) Moonquakes

(6) Instrumental effects

It is unclear whether any of the received signals were actually of lunar-seismic-event origin. One of the most important results of the PSEP is the discovery that the background noise level on the Moon is extremely low. At frequencies from 0.1 to 1 Hz, the background seismic-signal level for vertical surface motions is less than 0.3 mμm. This level is from 100 to 10 000 times less than average terrestrial background levels in the frequency range of 0.1 to 0.2 Hz for microseisms. Continuous seismic background signals from 10 to 30 mμm were observed on the records of the horizontal seismometers. These signals decreased considerably near lunar noon and may have been due to lunar surface-temperature changes which tilted the instrument.

Of the many seismic signals recorded, several were produced by the LM. Many of the signals may be a result of real seismic events and may be generated by moonquakes, impact events, or movement of surface rocks. However, none of the events can be clearly identified as real, and none of the observed signals has patterns normally observed on recordings of seismic activities occurring on Earth. Clearly, the Moon is not a very seismic body. Artificial seismic sources, such as the impact on the lunar surface of the Saturn IVB stage or the spent LM ascent stage, will be useful for future lunar seismometers.

Laser Ranging Retroreflector

The LRRR consists of an array of finely machined quartz corners deployed on the lunar surface and aimed at the Earth. The array is used as a reflector for terrestrial lasers. By measuring the distance from the laser to the reflector, small changes in the motion of the Moon or the Earth can be measured. The goal, when the system is fully operational, is an uncertainty of 15 cm (6 in.). The LRRR will allow studies to be conducted on (1) the librations of the Moon, both in latitude and longitude, (2) the recession of the Moon from the Earth caused either by tidal dissipation or by a possible change in the gravitation constant, and (3) the irregular motion of the Earth, including the Chandler wobble of the pole. The amplitude of the Chandler wobble seems to vary in time with relation to major earthquake events.

On the same day the LRRR was deployed satisfactorily on the surface of the Moon, attempts were made to range on the reflector from the Lick Observatory in California and from the McDonald Observatory in Texas. Some time was required for the ranging attempts to be successful because, initially, there was some uncertainty as to the location of the landing site. After a few days, this problem was solved, but ground-instrument difficulties and weather problems caused further delays. In approximately a week, both observatories had received signals reflected by the LRRR. The signals, although weak, were clearly identifiable. By using this technique, the distance to the Moon from the Earth has been measured to an accuracy of approximately 4 m. It should be noted that the distance to the Moon is actually uncertain to a few hundred meters because of uncertainties in the velocity of light. However, the distance expressed in light seconds is accurate to approximately one part in 10^9, which is consistent with the 4-m accuracy discussed previously.

The LRRR experiment will continue for months or years before final data are obtained on many of the detailed measurements to be undertaken.

The Solar-Wind Composition Experiment

For some time, direct measurements have been made of the solar wind, establishing the presence of approximately 5 percent helium ions in the solar-wind stream, which is composed predominantly of protons. The solar wind is expected to contain many heavier ions, probably representative of solar composition, but no direct measurements of these heavier species have yet been made. It now seems quite likely that the rare gas measured in the powdery lunar samples is of solar-wind origin, but this rare-gas source may be confused with other gas sources. The measured rare gas represents an integration of the solar wind into the soil over a period of many millions of years.

An experiment was conducted during the Apollo 11 mission to measure heavier elements in the solar wind directly. A thin aluminum foil of 4000 cm^2 was deployed on the surface of the Moon facing the Sun. The solar-wind particles were expected to penetrate approximately 10^{-5} cm into the foil and to be firmly trapped there. The foil was collected after 77 min, placed inside one of the sample-return containers, and brought to the NASA Manned Spacecraft Center (MSC) Lunar Receiving Laboratory (LRL).

Approximately 1 ft^2 of the foil was removed in the LRL, sterilized by heat at 125° C for 39 hr, and sent to Switzerland for analysis. Several small pieces of the foil, each approximately 10 cm^2, were cleaned by ultrasonic methods, and the noble gases were then extracted for analysis in a mass spectrometer. Helium, neon, and argon were found, and their isotopic composition was measured. The results correspond generally to solar abundances and are clearly nonterrestrial. More complete results will be presented when the major portion of the foil has been analyzed.

The Lunar Samples

Of the 22 kg of returned lunar material, 11 kg are rock fragments larger and 11 kg are smaller than 1 cm in size. This material may be divided into the following four groups:

Type A — fine-grained vesicular crystalline igneous rock.

Type B — medium-grained crystalline igneous rock.

Type C — breccia.

Type D — fines (less than 1 cm in size).

The crystalline rocks contain mineral assemblages, crystal sizes, and gas cavities, indicating that the rocks were crystallized from lavas or near-surface melts. It is uncertain whether the lavas were impact generated or of internal origin. Twenty crystalline rocks were found in the returned sample, 10 rocks of type A and 10 of type B. Individual rocks weighed up to 919 g. The type A rocks contain vesicles of 1 to 3 mm in diameter faced with brilliant crystals. (Most vesicles are spherical but some are ovate.) There are also irregular cavities, or vugs, into which crystals and other groundmass minerals project. The percentages of the minerals present in type A rocks are clinopyroxene, 53 percent; plagioclase, 27 percent; opaques (including abundant ilmenite, minor troilite, and native iron), 18 percent; and other translucent phases, and a minor amount of olivine, 2 percent. Except for the high content of opaques, which reflects the high iron and titanium content, the mineralogy and chemistry of the rocks resemble terrestrial olivine-bearing basalts.

The dark-brownish-gray speckled type B rock is granular in texture and generally resembles terrestrial microgabbros. The grain sizes are from 0.2 to 3 mm. The percentages of minerals in type B rocks are as follows: clinopyroxene, 46 percent; plagioclase, 31 percent; opaques (mainly ilmenite), 11 percent; cristobalite, 5 percent; and other minerals, 7 percent. No olivine is present in type B rocks. Other unidentified minerals have been found. The complete absence of hydrous mineral phases in type A or type B rocks and the presence of free iron places a low limit on the amount of water present in the melt from which the rocks crystallized. The water content in the lunar material is considerably lower than the water content in terrestrial basalts.

All the breccias are mixtures of fragments of different rock types similar to type A or type B rocks and are mixtures of angular fragments and sphericles of glass of a variety of colors and a variety of indices of refraction. Evidence of strong shock is present in many type C rocks. These rocks vary from being very friable and soft to being as hard as the crystalline rocks. Evidence points to the breccias being formed by shock cementing or by lithification of the powdery lunar surface material. Both techniques of breccia formation are probably the result of impact events.

Fines were returned to Earth in the bulk-sample container and in the core tubes. The core tubes showed no stratification of fines, and the material was, in general, similar to the bulk-sample fines. The type D material consists of a variety of glasses, plagioclase, clinopyroxene, ilmenite, and olivine. A few Ni-Fe spherules up to 1 mm in diameter were found. Glasses, which make up approximately 50 percent of type D material, are of three types: (1) vesicular, globular dark-gray fragments; (2) pale or colorless angular fragments with refraction indices of approximately 1.5 to 1.6; and (3) spheroidal, ellipsoidal, dumbbell-shaped, and teardrop-shaped bodies. Most of these glass types are smaller than 0.2 mm and range in color from red to brown to green to yellow with refraction indices ranging from 1.55 to 1.8. Unlike most terrestrial magnetic glasses, many single glass particles are inhomogeneous. Like the Ni-Fe spherules, the type 3 rounded glass bodies seem to indicate melting induced by strong shock. In fact, evidence for shock is common in the fines and in the breccias. However, only a few crystalline rocks show evidence of strong shock in places other than near the surface pits.

A chemical analysis of several samples of all four types of glass was made using optical emission spectroscopy. A few samples were also analyzed using atomic absorption procedures. All four types of lunar material samples have been studied, and all types appear to be quite similar chemically. There is no similar material common on Earth. Several of the refractory metals are very prominent. On Earth, a deposit containing from 5 to 10 percent titanium is rare and might be considered a titanium mine. Zirconium, yttrium, and chromium are also present in amounts substantially larger than might be expected in terrestrial basalts.

Another characteristic of the samples is that they are low in alkali and volatile elements such as sodium, potassium, rubidium, lead, and bismuth. Aside from this characteristic and the very low water content mentioned previously, the material might be considered to be similar to terrestrial basalts. Because of the enrichment of the refractory elements and the depletion of the volatile elements, it is tempting to consider the material to be similar to a cinder, being the end product of a high-temperature environment, to explain these modifications.

The rare-gas content of lunar samples has been measured by mass spectroscopy. The content of beryllium, neon, argon, krypton, and xenon has been measured in several samples. The total rare-gas content in the fines and breccias is approximately 0.1 cc/g, which is quite high. These gases were found in the outside layer of the crystalline rocks, but the gas content found inside these rocks was very low. The gases were found throughout the breccias, further indicating that the breccias are formed, probably by shock, from the fines. Quite clearly, most of the measured rare gas is from the solar wind. Although no isotopic ratios have been measured previously for solar rare gases, the values obtained from the preliminary examination compare well with the values that would theoretically be expected.

Several of the rocks measured contained radiogenic ^{40}Ar, which, when coupled with data on potassium abundance, enables dating of the time of crystallization of the rocks by using the decay process

$$^{40}Ar \longrightarrow {}^{40}K + e^+ \qquad (1)$$

This process gives an age of $3.0\pm0.7\times10^9$ yr, which shows that the surface of the maria is older than had been expected.

Age dating can also be accomplished by measuring the amount of certain isotopes that are produced by cosmic ray bombardment of the lunar surface. This technique yields the time that the rocks within a few feet of the lunar surface have been exposed to cosmic rays. The rocks studied had cosmic-ray-exposure ages from 20×10^6 to 160×10^6 yr.

The organic content of the samples was studied by two techniques: (1) pyrolysis with flame ionization detection and (2) mass spectroscopy. The amount of extractable organic material with temperatures to 500° C is less than 1 part per million (ppm). Some organic contaminants were introduced into the samples by recovery and processing in the LRL, and it is uncertain whether an indigenous organic material exists in the samples.

Several samples have been counted in the LRL Radiation Counting Laboratory. The gamma rays measured show the presence of several cosmic-ray-induced radionuclides and the presence of thorium, uranium, and potassium. The uranium and thorium concentrations are near the concentration values of these elements found in terrestrial basalts; however, the amount of potassium present in the lunar samples is lower than that in terrestrial basalts and is comparable to the amount found in chondritic meteorites.

Conclusions

The major findings have been as follows:

(1) By using a fabric and mineralogy basis, the rocks can be divided into two genetic groups: fine- and medium-grained crystalline rocks of igneous origin and breccias of complex origin.

(2) The crystalline rocks, as shown by their modal mineralogy and bulk chemistry, are different from terrestrial rock and meteorites.

(3) Erosion has occurred on the lunar surface in view of the fact that most of the lunar rocks are rounded, and some of the rocks have been exposed to a process that gives them a surface appearance similar to sandblasted rocks. There is no evidence of erosion by surface water.

(4) The probable presence of the assemblage iron-troilite-ilmenite and the absence of any hydrated phase indicate that the crystalline rocks were formed under extremely low partial pressures of oxygen, water, and sulfur. (Pressures are in the range of equilibrium pressures found within most meteorites.)

(5) The absence of hydrated minerals suggests that no surface water has existed at Tranquility Base at any time since the rocks in this region were exposed.

(6) Evidence of shock or impact metamorphism is common in the lunar rocks and fines.

(7) All the rocks have glass-lined surface pits that may have been caused by the impact of small particles.

(8) The fine material and the breccia contain large amounts of all the noble gases which have elemental and isotopic abundances indicative of origin in the solar wind. The fact that interior samples of the breccias contain these gases implies that the samples were formed at the lunar surface from material previously exposed to the solar wind.

(9) $^{40}K - ^{40}Ar$ measurements on igneous rocks show that they crystallized from 3×10^9 to 4×10^9 yr ago. The presence of nuclides produced by cosmic rays shows that the rocks have been within 1 m of the surface for periods of 20×10^6 to 160×10^6 yr.

(10) The level of indigenous organic material capable of volatilization or pyrolysis, or both, appears to be extremely low, that is, considerably less than 1 ppm.

(11) Chemical analyses of 23 lunar samples show that all rocks and fines generally are chemically similar.

(12) The elemental constituents of lunar samples are the same as those found in terrestrial igneous rocks and meteorites. However, the following significant differences in composition were observed: some refractory elements (for example, titanium and zirconium) are notably enriched, and the alkali and some volatile elements are depleted.

(13) Elements that are enriched in iron meteorites (i.e., nickel, cobalt, and the platinum group) were not observed; such elements are probably low in abundance.

(14) Of 12 radioactive species identified, two were cosmogenic radionuclides of short half life, namely ^{52}Mn (5.7 days) and ^{48}V (16.1 days).

(15) Uranium and thorium concentrations are near the typical concentrations of these elements found in terrestrial basalts; however, the potassium-to-uranium ratio determined for lunar surface material is much lower than such ratios determined for terrestrial rocks or meteorites.

(16) No evidence of biological material has

been found to date in the lunar samples.

(17) The lunar soil at the landing site is predominantly fine grained, granular, slightly cohesive, and incompressible. The hardness of the lunar soil increases considerably at a depth of 15 cm. The soil is similar in appearance and behavior to the soil encountered at the Surveyor equatorial landing sites.

(18) The PSEP deployed on the Moon operated satisfactorily for 21 days and detected several seismic signals, many of which originated from astronaut activity or mechanical motions of the LM. It is uncertain whether any actual lunar seismic events were detected. The seismic-noise background is much less on the Moon than on Earth.

(19) The LRRR was deployed on the Moon and has been used as a target for Earth-based lasers. The distance to the Moon has now been measured to within an accuracy of approximately 4 m. Future studies will be made on the variation of this distance to study in detail the motion of the Moon and the Earth.

(20) Preliminary analysis has been made on part of the aluminum SWC foil. Helium, neon, and argon have been found in the analysis, and the isotopic composition of each element has been measured.

1. Photographic Summary of Apollo 11 Mission

James H. Sasser

The geographical exploration of new frontiers has usually occurred many years before scientists visited and studied the areas in detail. For example, the existence of Antarctica as a continent was known from the time Charles Wilkes explored 1500 miles of the coastline in 1840. However, extensive Antarctic exploration did not begin until the 20th century with the voyages of Scott, Amundsen, Shackleton, and Byrd. It was not until the International Geophysical Year (July 1957 to December 1958) that scientists from 12 countries began conducting an ambitious Antarctic research program. In this respect, the first manned lunar exploration was unique. The scientific experiments were carefully planned, and the astronauts were trained as surrogates for scientists representing many disciplines.

A brief description of the Apollo 11 mission, illustrated with photographs taken by the astronauts during the mission, is presented. All photography in this chapter was taken on 70-mm film with Hasselblad cameras except figure 1. The cameras have a motor-driven mechanism powered by two nickel-cadmium batteries to advance the film and cock the shutter after each picture is taken. The photographs taken from the command module (CM) were made with 80- or 250-mm lenses. The lunar surface photography was taken with 80- and 60-mm lenses. The small crosses seen in the lunar surface photography are reseau marks that are used for calibration purposes during postflight data reduction. A 16-mm sequence camera for data acquisition was used to document many phases of the mission.

During the mission, 9 magazines of 70-mm film and 13 magazines of 16-mm film were exposed. The black and white photography was taken on type 3400 Panatomic-X film emulsion on a 2.5-mil Estar polyester base. Ektachrome EF SO168 color film on a 2.5-mil Estar polyester base was exposed on the lunar surface. The higher speed of this color film was expected to be more suitable for lunar surface photography because of the low light levels anticipated and confirmed to exist on the lunar surface. Other color 70-mm exposures of the Earth and Moon were taken on Ektachrome MS SO368 color reversal film on a 2.5-mil Estar polyester base.

The 16-mm film taken during lunar module (LM) descent provided the first accurate knowledge of the exact landing point on the lunar surface. The 70-mm photographs taken on the lunar surface provided panoramic views of the surface near the landed LM and allowed detailed topographic mapping of the lunar surface near the landing point.

The photographs shown are only a sample of those obtained. Study of the photography will continue for many years.

Launch, Earth Orbit, and Translunar Coast

The Apollo 11 spacecraft was launched at 13:32:01 G.m.t. on July 16, 1969, from Cape Kennedy, Florida. All times indicated later in this chapter are stated in hours and minutes from launch time (ground elapsed time (g.e.t.)), to the nearest minute. At 3 min after launch, Saturn V first-stage (S-IC) engine cutoff and second-stage (S-II) engine ignition occurred, and the launch escape tower was jettisoned. At 9 min after launch, S-II engine cutoff and third-stage (S-IVB) engine ignition took place. Earth-orbit insertion of the spacecraft was achieved at 12 min after launch.

After 2 hr and 33 min in Earth orbit, the S-IVB engine was reignited for acceleration of the spacecraft to the velocity required for Earth-

gravity escape. At 3:17 g.e.t., the command and service module (CSM) was separated from the S-IVB, turned around, and docked with the LM. The LM was removed from the S-IVB at 4:17 g.e.t.

Provision was made in the flight plan for four midcourse corrections (using the service module propulsion system to refine the spacecraft trajectory); however, the only necessary midcourse correction took place at 26:45 g.e.t. Lunar-orbit insertion of the spacecraft began at 75:50 g.e.t.

FIGURE 1-1. — On July 16, 1969, at 9:32 A.M. e.d.t., 7½ million lb of thrust lifted the Apollo-Saturn V spacecraft carrying Astronauts Neil A. Armstrong, Michael Collins, and Edwin E. Aldrin, Jr., from the launch site at Cape Kennedy.

FIGURE 1-2. — Shortly after translunar insertion, the astronauts photographed Tropical Storm Bernice in the western Pacific Ocean off the coast of Baja California. (NASA AS11-36-5298)

FIGURE 1-3. — Compare this westward view (NASA AS11-36-5300) of the central portion of Baja California, taken shortly after the spacecraft left Earth orbit on a trajectory to the Moon, with the next picture.

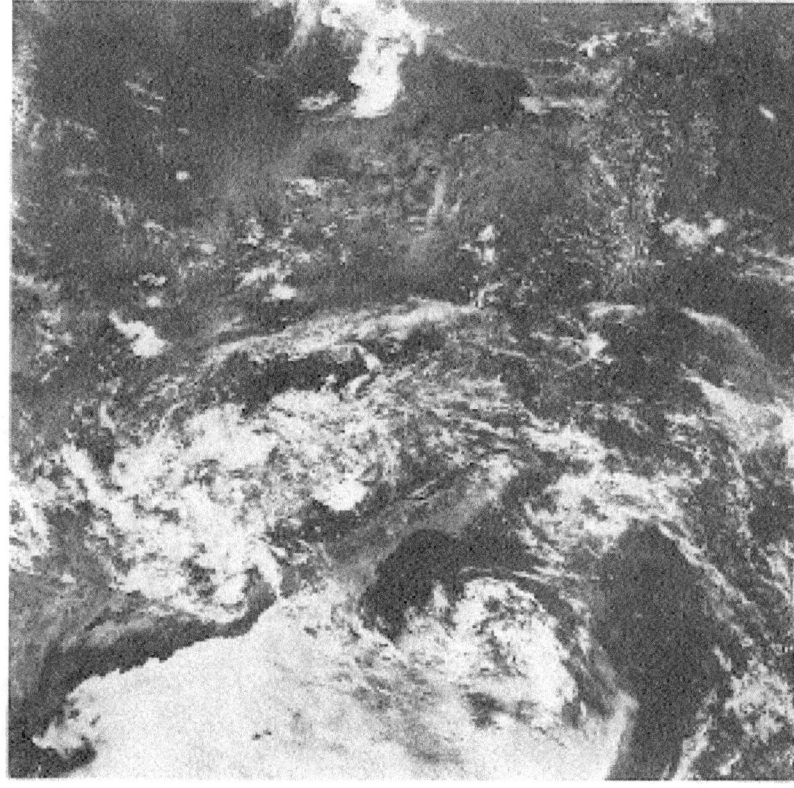

FIGURE 1-4. — This view (NASA AS11-36-5308) of the southern portion of the North American Continent shows Baja California again, the coast of southern California, and the mountains of Mexico.

FIGURE 1-5. — Separation of the CSM from the S-IVB occurred at 3:17 g.e.t. The panels that protected the LM during launch were ejected, and the CSM was docked with the LM at 3:25 g.e.t. This photograph (NASA AS11-36-5313) was taken shortly before docking.

FIGURE 1-6. — This photograph (NASA AS11-36-5355) of the Earth, taken from a distance of more than 100 000 n.mi., was made approximately 25 hr after launch. The Eastern Hemisphere was remarkably cloud-free.

Lunar Orbit

The lunar-orbit insertion maneuver was begun at 75:50 g.e.t. The spacecraft was placed in an elliptical orbit (61 by 169 n.mi.), inclined 1.25° to the lunar equatorial plane. At 80:12 g.e.t., the service module propulsion system was reignited, and the orbit was made nearly circular (66 by 54 n.mi.) above the surface of the Moon. During each 2-hr orbit, the terminator (the line between daylight and darkness) moved westward across the Moon at the rate of approximately ½ deg per hour (30 km per orbit).

During the 59 hr and 34 min of spacecraft lunar orbit, approximately 210° of the lunar surface below the spacecraft orbit was illuminated by the Sun and could be photographed. At the orbital altitude, the lunar horizon was approximately 300 n.mi. from the spacecraft nadir.

Photographs taken from lunar orbit provide synoptic views for the study of regional lunar geology. The photographs are used for lunar mapping and geodetic studies, and they have been valuable in training the astronauts for future lunar missions. The following selected photographs, taken from lunar orbit, are arranged in geographical sequence from the farside terminator to the nearside terminator, without regard to the times at which they were taken. The regions of the Moon covered by the selected photographs are plotted on the map that follows.

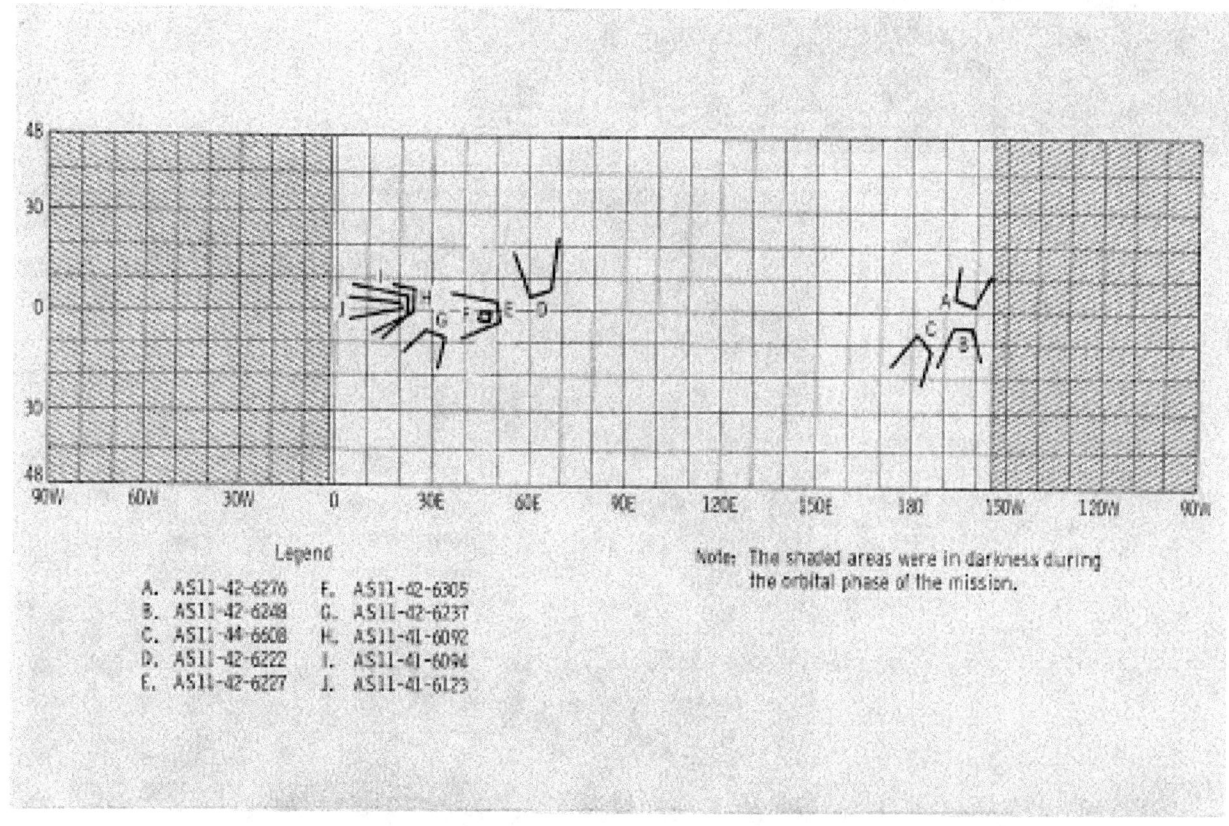

FIGURE 1-7. — This diagram shows the approximate areal coverage plot of selected orbital photographs.

FIGURE 1-8. — The view in this photograph (NASA AS11–42–6276) is toward the northeast near the farside terminator. The view is centered at 10° N, 163° W. The features shown are not named.

FIGURE 1-9. — The far side of the Moon is pictured, looking south. The features included in this view have not yet been named. The coordinates of the center of this photograph (NASA AS11–42–6248) are approximately 8° S, 164° W.

FIGURE 1-10. — The center of this photograph (NASA AS11-44-6608) is located approximately 10° S, 177° W, and shows the view to the southwest from the spacecraft. Note the smooth mare-type fill in the valleys in the foreground.

FIGURE 1-11. — The view looks northwest across Mare Crisium (the Sea of Crises). The large, dark-floored crater Firmicus is in the foreground. The average elevation of the highlands is several thousand meters above the floor of the mare. (NASA AS11-42-6222)

FIGURE 1-12. — The craters Messier, Messier A, and Messier B are grouped in the center of this photograph (NASA AS11-42-6227), looking westward across Mare Fecunditatis. Messier appears nearest the center of the photograph, Messier A is above Messier, and Messier B is below and to the right of Messier.

FIGURE 1-13. — This is a telephoto view of the craters Messier and Messier A. The atypical shape of these craters has caused considerable controversy concerning the mode of origin. (NASA AS11-42-6305)

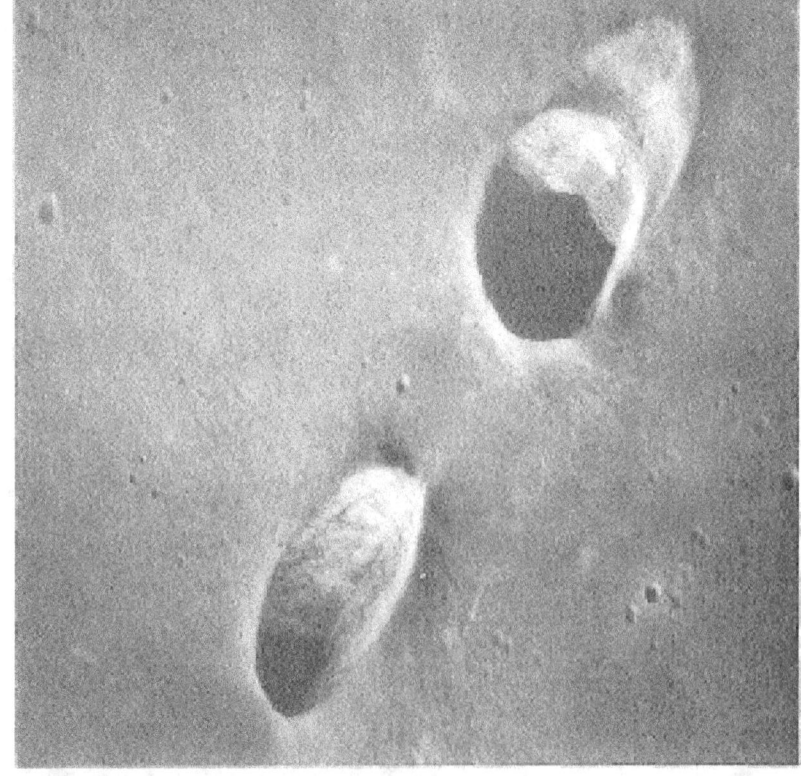

FIGURE 1-14. — This view is looking southward across the crater Theophilus, which is approximately 60 miles in diameter. To the east of Theophilus is Madler, a crater approximately 14 miles in diameter. The center of the view is located approximately 11° S, 29° E. (NASA AS11-42-6237)

FIGURE 1-15. — The twin craters Sabine and Ritter lie beyond the landing site. (NASA AS11-41-6092)

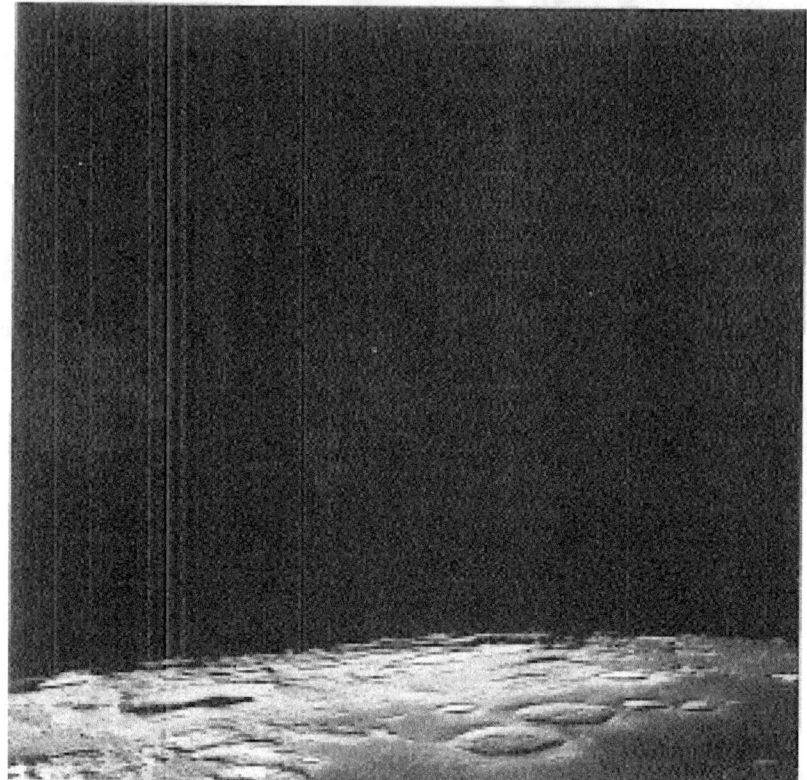

FIGURE 1-16. — Beyond the landing site, looking toward the sunrise terminator on the Moon, the craters Sabine and Ritter lead into the highlands between the Sea of Tranquility (Mare Tranquillitatis) and the Central Bay (Sinus Medii). (NASA AS11-41-6094)

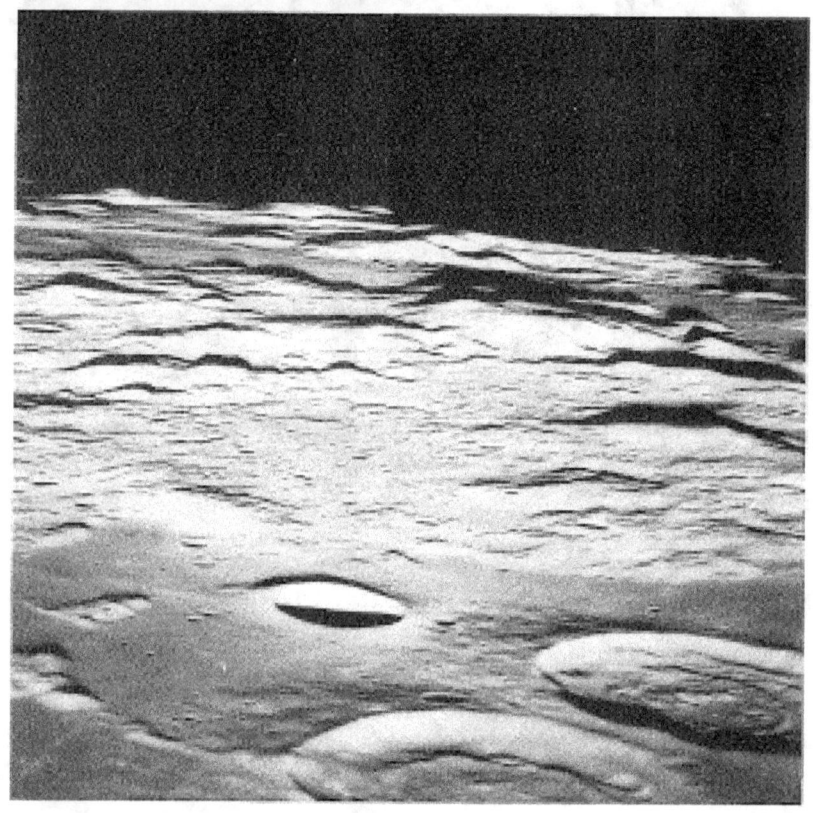

FIGURE 1-17. — The lunar highlands west of the Sea of Tranquility stand out in stark relief when illuminated by the morning Sun. The twin craters Sabine and Ritter are in the right foreground, and the crater Schmidt is in the central foreground. (NASA AS11-41-6123)

Lunar Module Descent and Landing

The LM, with Astronauts Armstrong and Aldrin aboard, was undocked from the CSM at 100:14 g.e.t., following a thorough check of all the LM systems. The CSM (radio call sign "Columbia") was maneuvered away from the LM (radio call sign "Eagle"). At 101:36 g.e.t., the LM descent engine was fired for approximately 29 sec, and the descent to the lunar surface began. At 102:33 g.e.t., the LM descent engine was started for the last time and burned until touchdown on the lunar surface. Eagle landed on the Moon 102 hr, 45 min, and 40 sec after launch.

FIGURE 1-18. — Eagle was photographed from Columbia by Astronaut Collins at the moment the two spacecraft undocked. (NASA AS11-44-6586)

Figure 1-19. — Shortly after undocking, Eagle was rotated to permit a visual inspection by Astronaut Collins in Columbia. The probes that extend downward from three of the LM footpads sensed contact with the surface during touchdown. (NASA AS11-44-6584)

Figure 1-20. — A remarkable sequence of three photographs, taken from Eagle. (a) An oblique view (NASA AS11-37-5437) of the LM landing site.

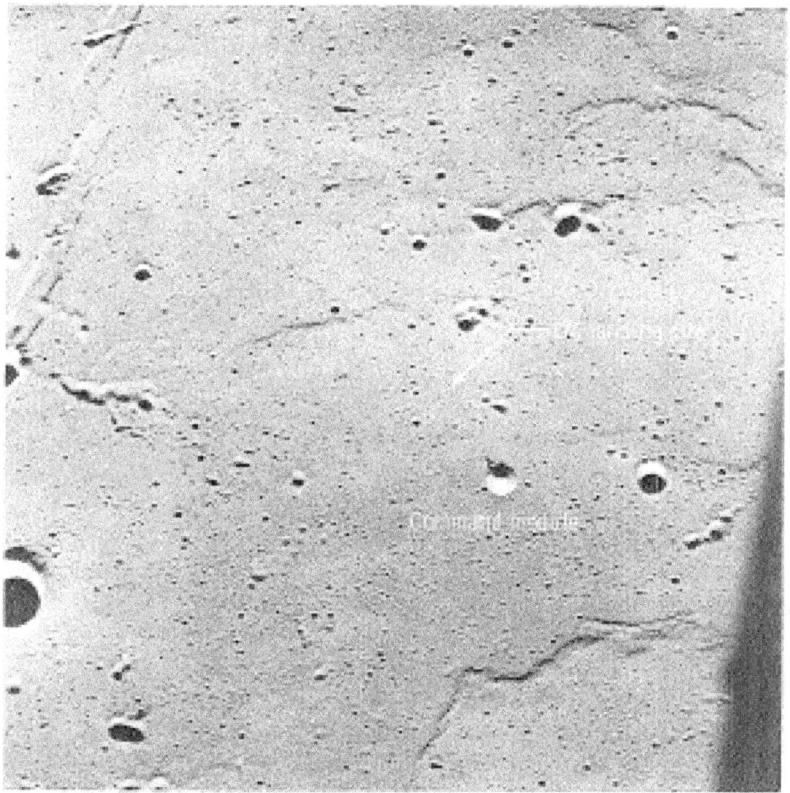

Figure 1-20(b).—An almost vertical view (NASA AS11-37-5447) of Columbia over the LM landing site.

Figure 1-20(c).—A vertical view (NASA AS11-37-5448) of Columbia almost 60 n.mi. above the crater Schmidt as the two spacecraft approach the terminator.

Lunar Surface Activities

Immediately after landing on the Moon, Astronauts Armstrong and Aldrin prepared the LM for liftoff as a contingency measure. A series of photographs was taken through the LM windows during this activity; and when the simulated liftoff countdown was completed, the astronauts ate. Following the meal, a scheduled sleep period was postponed at the astronauts' request, and the astronauts began preparations for descent to the lunar surface.

Astronaut Armstrong emerged from the spacecraft first. While descending, he released the Modularized Equipment Stowage Assembly (MESA) on which the surface television camera was stowed, and the camera recorded man's first step on the Moon at 109:24:19 g.e.t. A sample of the lunar surface material was collected and stowed to assure that, if a contingency required an early end to the planned surface activities, samples of lunar surface material would be returned to Earth. Astronaut Aldrin subsequently descended to the lunar surface.

The astronauts carried out the planned sequence of activities that included deployment of a Solar Wind Composition (SWC) experiment, collection of a larger sample of lunar surface material, panoramic photographs of the region near the landing site and the lunar horizon, closeup photographs of inplace lunar surface material, deployment of a Laser-Ranging Retroreflector (LRRR) and a Passive Seismic Experiment Package (PSEP), and collection of two core-tube samples of the lunar surface. Approximately 2¼ hours after descending to the surface, the astronauts began preparations to reenter the LM, after which the astronauts slept. The ascent from the lunar surface began at 124:22 g.e.t., 21 hr and 36 min after the lunar landing.

The photographs reproduced in this section consist of mosaics (photographs joined together to form panoramic views) and single photographs (to show more detail in specific areas). The single photographs are arranged in the sequence in which they were taken.

Before EVA

After EVA

FIGURE 1-21. — Apollo 11 lunar surface panoramas taken from the LM.

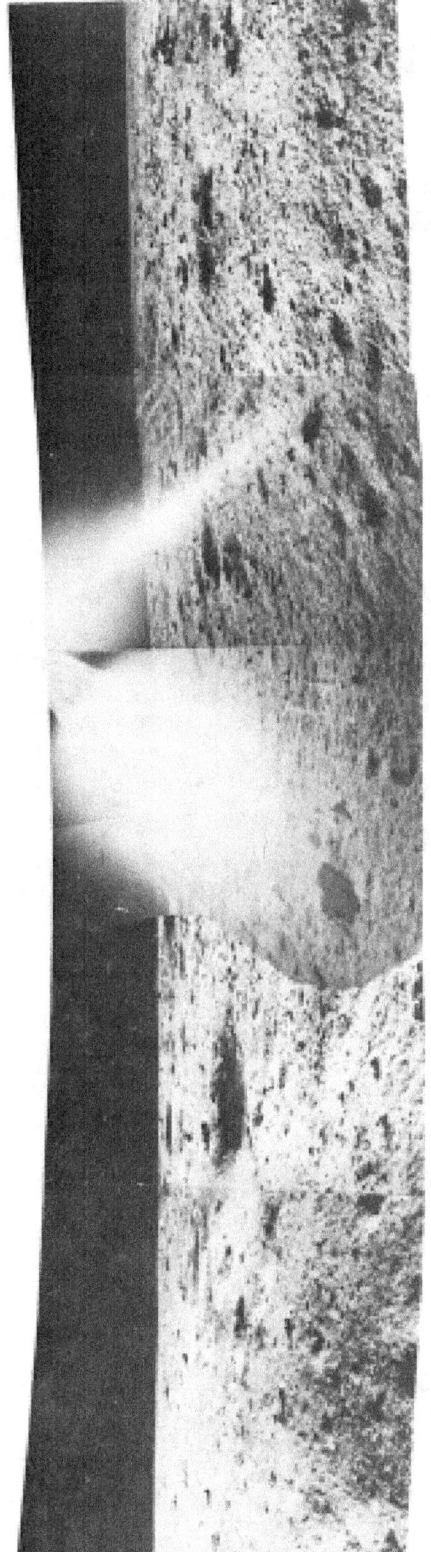

Panorama of the lunar surface looking east

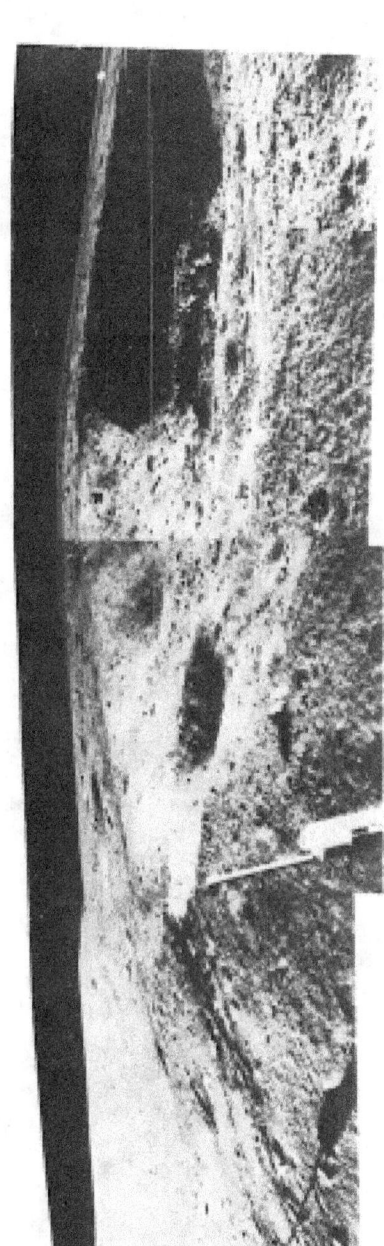

Panorama of the large crater approximately 200 ft east of the LM

FIGURE 1-22. — Apollo 11 panoramas taken from the surface.

Panorama of the lunar surface looking northwest

Panorama of the lunar surface looking south

Panorama of the lunar surface looking north

FIGURE 1-23. — Additional panoramas from the lunar surface.

FIGURE 1-24.— Astronaut Aldrin descends the ladder to the lunar surface. (NASA AS11-40-5868)

FIGURE 1-25.— The SWC experiment is unrolled and turned to face the Sun. The experiment was deployed on the surface for approximately 1 hr and 17 min. The linear trails from the foreground toward the LM were formed by the cable of the surface television camera. (NASA AS11-40-5872)

FIGURE 1-26.— Note the rounded, eroded appearance of the rock in the right foreground. (NASA AS11-40-5875)

FIGURE 1-27.— The astronaut photographed his own footprint to permit later study of the lunar surface bearing strength. The thin, crusty appearance of the surface was similar to that discovered during the Surveyor soil mechanics experiments. (NASA AS11-40-5877)

FIGURE 1-28.— The LM footpads were also photographed for later study of the surface mechanical properties. The foil wrapping provides thermal insulation. (NASA AS11-40-5926)

FIGURE 1-29.— The LRRR and the PSEP are being removed from the MESA in the LM descent stage. The Apollo Lunar Surface Closeup Camera (ALSCC), which provides stereoscopic pictures of the fine surface structure, is in the foreground. The television camera can be seen beyond and to the right of the SWC. (NASA AS11-40-5931)

Figure 1-30.—The LRRR and the PSEP are carried to the deployment area. The surface was softer near the rim of the small crater in the foreground, as can be seen by the depth of Astronaut Aldrin's footprints. (NASA AS11-40-5944)

Figure 1-31.—The LRRR has been set up to face the Earth, and the PSEP is being leveled. The PSEP was placed behind the large rock to shield the experiment from the effects of liftoff. (NASA AS11-40-5949)

Figure 1-32.—Shortly before entering the LM, Astronaut Armstrong walked back approximately 200 ft eastward to photograph the interior of a crater he noted during descent. The ALSCC is in the foreground. (NASA AS11-40-5957)

Figure 1-33.—Tranquility Base is shown from the rim of the crater shown in the previous photograph. Astronaut Armstrong's shadow is in the left foreground, and the shadow of the ALSCC is in the right foreground. (NASA AS11-40-5961)

FIGURE 1-34. — The astronaut is collecting a core-tube sample for later study by scientists. The SWC experiment is still deployed. (NASA AS11-40-5964)

The Return to Earth

At 124:22 g.e.t., the engine of the LM ascent stage was ignited, and a series of maneuvers was begun to permit rendezvous with Astronaut Collins in the CM. Docking took place at 128:03 g.e.t. Astronauts Armstrong and Aldrin transferred to the CM, and the LM ascent stage was jettisoned at 130:09 g.e.t. The service module (SM) engine was reignited at 135:24 g.e.t. to increase the spacecraft velocity by 3279 fps for escape from lunar orbit.

In transearth coast, as in translunar coast, only one of four planned midcourse corrections was required. This midcourse correction was performed at 150:30 g.e.t. The SM was separated from the CM at 194:49 g.e.t. The CM entered the atmosphere of the Earth with a velocity of 36 194 fps and landed in the Pacific Ocean 195 hr, 18 min, and 35 sec after launch. Procedures to prevent biological back-contamination of the Earth were followed in recovering the crew, the lunar samples, and the spacecraft.

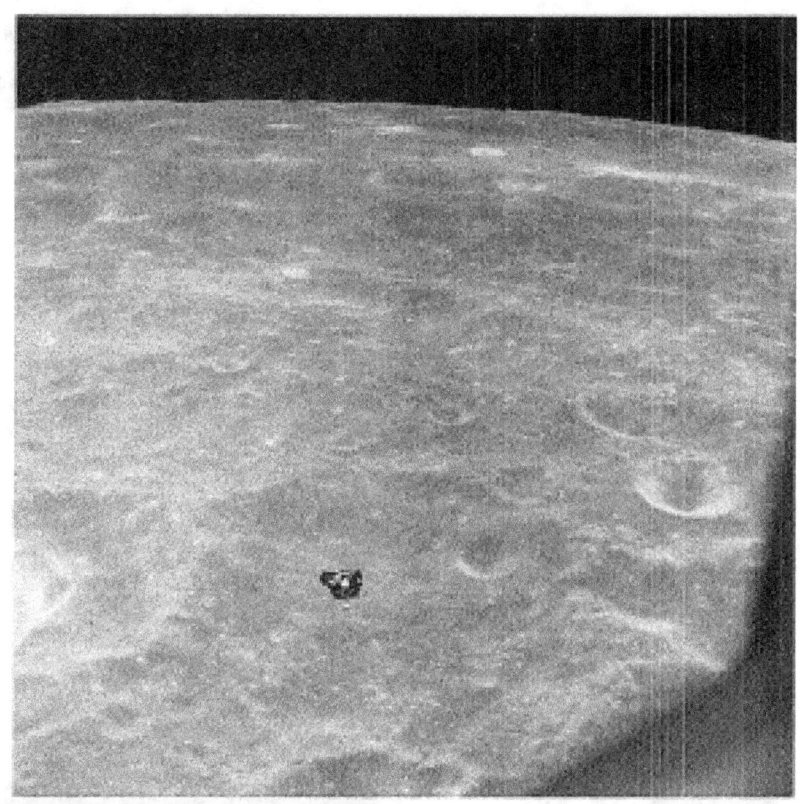

FIGURE 1-35.—The lonely vigil of Astronaut Collins in the CM nears an end with the approach of Astronauts Armstrong and Aldrin in the ascent stage of the LM. The center of the view is at lunar coordinates 1.5° N, 105° E. (NASA AS11-44-6621)

FIGURE 1-36.—Near 88° E, the two spacecraft approach Mare Smythii as the Earth rises above the lunar horizon. (NASA AS11-44-6642)

FIGURE 1-37.— A nearly full Moon is viewed from a perspective never seen from Earth, approximately 60° E. The LM landing site is far to the west in this view. (NASA AS11-44-6665)

FIGURE 1-38.— A crescent Earth awaits the return of the first men to set foot on the Moon. (NASA AS11-44-6889)

2. Crew Observations

Edwin E. Aldrin, Jr., Neil A. Armstrong, and Michael Collins

The Apollo 11 mission offered us our first opportunity to observe directly the operational and scientific phenomena associated with landings on the lunar surface. The lunar module (LM) crew participated in two mission-oriented geologic field trips and also attended several tens of hours of briefings dealing with the observational aspects of the mission. These briefings included sessions on basic geologic principles, rock and mineral identification and interpretation, impact geology, Surveyor program results, lunar surface visibility characteristics, landing site recognition, and LM landing characteristics. In addition, considerable planning and simulation time was devoted to preparation for the evaluation of extravehicular mobility, a vital feature to scientific endeavors on the lunar surface.

Landing and Ascent

We had trained to some extent on the specific topographic characteristics of the site 2 landing area to be able to identify our final landing point, an important requirement for later missions. Operational problems during the early visibility phase of final descent prevented observation of those features that would have enabled us to locate our final landing point soon after touchdown. Postmission examination of the descent sequence photography showed several features we had been trained to recognize; therefore, we feel that it will be feasible to locate landing points on future missions.

When we were first able to observe the terrain ahead of us (at an altitude of about 2000 ft and about 1 mile uprange), it was clear that we would land in the vicinity of a blocky rim crater about 200 yd in diameter. This crater was later identified as one we had informally called West Crater during our prelaunch training. The blocky area just north of West Crater was initially considered for a landing site because of its obvious scientific interest; but as we approached this area, it proved to be too rough for landing because of the numerous large and blocky rocks, many larger than 5 or 10 ft in size. We elected to overfly this area in preference for the smoother spots a few hundred yards farther west. As we proceeded west, it was noted that several relatively open areas existed between raylike fields of boulders. Our final approach was chosen to lie between two of these boulder fields.

The site where we eventually landed was a relatively smooth area; however, craters varying in diameter from 1 to perhaps 50 ft were present in the near vicinity of the touchdown point. Crater density in this area appeared to be inversely proportional to crater size; for example, the small-diameter craters were much more abundant than those having a large diameter. Several subdued ridges 20 to 30 ft high were within our field of view but were some distance from the spacecraft. In addition, a long boulder field with many boulders averaging at least 2 ft in diameter was several hundred feet north of the landing point.

A large amount of fine lunar material appeared to be moved during the terminal phase of landing. The streaming dust was first evident at about 100 ft in altitude and thickened to about 75 percent loss of visibility at touchdown. The path of this expelled material was parallel to the lunar surface, and the material seemed to travel a great distance (over the horizon) at a high velocity. The dominant effect of the streaming dust was to partially obscure and distort the details of the surface and to make it difficult to judge the horizontal velocity of the LM. After the engine was shut down, there was no visual evidence that the expelled material was deposited anywhere on top of the surrounding surface, and no lingering cloud of dust was observed. No lunar dust was observed later during liftoff; however, separation debris observed at this time appeared to have a long flightpath similar to the path exhibited by the lunar material that was expelled during the terminal phase of landing. One piece of debris was observed to follow below us for several miles before impacting the surface.

General Surface Characteristics

Color

The general color of the surface as viewed close at hand is comparable to that we observed from orbit at the same Sun elevation of about 10°. Actually, the surface is pretty much without color other than shades of gray. Along the zero phase angle line (down Sun), the color is a light, chalky gray with a tan shade. As one looks toward 90° to the sunline, the color becomes a darker ashen gray with no apparent tan shades. At all phase angles, the surfaces of rock fragments appear to be a lighter gray than their immediate surroundings; however, where such fragments are freshly broken, the broken surfaces are very dark gray, much like country basalt. Small gradations in color resulting from very small topographical changes were visible. Upon looking closely at the surface material (particularly the fine soil) during our excursion, we observed a charcoal-gray to gray-cocoa color, similar to that of a graded lead pencil.

Craters

Many, if not most, of the larger craters in the immediate vicinity of the LM appeared to be somewhat elongate secondary craters. Except for a few of the smaller craters, almost all the craters we observed had raised rims; however, no rims were truly sharp. Other than West Crater, we did not observe a crater whose rim had a noticeably greater abundance of rock fragments than its surroundings had. Also, there were no observed indications of fragments having rolled down the sides of craters.

The floors of craters near Tranquility Base generally have about the same fragment concentration as the rims have. The only exception to this was the 33-ft-diameter, 15- to 20-ft-deep crater about 200 ft east of the LM. This crater had a mound of large rocks grouped slightly off center that were much larger and more numerous on the bottom than on the wall or rim of the crater.

Rock Fragments

Rock fragments of a wide range of size, angularity, and textural types were observed both from the LM window and during our extravehicular activities. Although some rock fragments were obviously lying on top of the surface, it was not always possible to judge their depth of burial. In one case, a relatively angular fragment resembling a distributor cap was found to have most of its volume extending about a foot underground, with a flat shelf at ground level around the exposed part. In another instance, a rock about 1 ft high, 1½ ft long, and ½ ft thick was apparently standing on edge, although its depth of burial was unknown. In the course of our use of the scoop, we encountered rocks buried under several inches of soil.

The major textural types of the rock fragments we observed were as follows: plain, even-grained basaltlike rocks; obviously vesicular basaltlike rocks; basaltic-appearing rocks with 1 to 5 percent small white minerals; and rocks consisting of aggregated smaller fragments. In some instances, loosely aggregated clods of soil were difficult to distinguish from rock fragments until they were disturbed and broken up.

Some of the rocks that appeared to be vesicular at first glance were actually not vesicular. Instead, they were surficially pitted by what appeared to be small impact craters.

In addition to the rock fragments, we observed smaller pieces of material that had a metallic luster and resembled blobs of solder splattered on an irregular surface. These pieces of material were concentrated in scattered aggregates at the bottoms of six or eight 3- or 4-ft-diameter craters. Elsewhere, several examples of lunar material that seemed to be transparent crystals were observed on the surface. These crystallike materials resembled quartz crystals and appeared to be opaque from some views and translucent from other views. There were also small sparkly fragments that resembled biotite; however, these fragments could not be examined closely. It was not possible to collect samples of these unusual materials at the time they were observed, and during the sampling periods, none of these materials were found. The solderlike material was photographed, however, using the closeup stereoscopic camera.

Soil Characteristics

The fine surface material is a powdery, graphitelike substance that seems to be dominantly sand to silt size. When this material is in contact with rock, it makes the rock slippery. This phenomenon was checked on a fairly smooth, sloped rock. When the powdery material was placed on the rocks, the boot sole slipped easily on the rock, and the slipping was sufficient to cause some instability of movement. Otherwise, traction was generally good in the loose, powdery material.

Unexpected differences in the consistency and softness of this top layer were noticed at locations having minor changes in surface topography. These differences were manifested in significantly different footprint depths. These depth differences indicated that there may be different depths of surface material covering the more resistive subsurface, particularly on the rims of small craters.

When the surface near the LM was viewed following touchdown, the topography that had been observed prior to touchdown had changed little. Although there was no crater, the lunar surface directly under the engine bell had a singed and baked appearance. The baking appeared to have made the surface material more cohesive and perhaps made it flow together to give a strong streaked pattern radiating from below the bell. However, the impact of small objects thrown on this material indicated that the material was about as loosely bound as the material away from the LM. The streaks, or ray texture, when viewed down Sun were much darker than the rest of the surface. Any erosion that existed in this area was clearly radially outward. However, this erosion was not obvious to us beyond the limits of the landing gear. No redeposition of material was observed away from the LM except possibly for the powder observed in the cavities of some rock fragments close to the spacecraft. Also, the soil under the LM showed no evidence of disruptive outgassing of injected engine gases.

Visual examination of the LM and of the lunar surface beneath the LM showed no effects from fluids that had vented from the LM. A slight amount of vapor that may have escaped from a boiler was observed at the top of the LM.

Surface penetrability decreases quickly within the first few inches of the surface. The first evidence of this decrease was the 2- to 3-in. penetration of the LM footpads and the rather shallow trenching by the landing probes. When specifically probed more than 4 or 5 in., the surface was found to be quite firm. This firmness was clearly evident during deployment of the staffs of the U.S. flag and the Solar Wind Composition Experiment. Probing of the initial 4 or 5 in. of the surface was relatively easy; however, 6 to 8 in. was as far as the flagstaff would penetrate the surface. The surface penetration by the core tubes was no greater than 8 or 9 in., even when the sampler extension was hammered hard enough to be significantly dented. In contrast, the soil offered very little lateral support to the staffs and core tubes when they were left to stand by themselves. There were no rocks under the core tubes during the driving operations. The material at the bottom of the core sample appeared to be darker than the surface material, and this material packed in the tube and adhered to the sides of the tube in the same manner as wet sand or silt.

When kicked or scuffed, lunar material moved away radially. The ballistic trajectory patterns of this kicked material depended upon the angle of impact of the boot. Most of the particles had the same angle of departure and velocity so that a large proportion of the particles impacted the same distance from their source. In the lunar environment, a cloud of kicked material did not settle slowly as would have occurred if material were kicked on Earth. The material disturbed by our walking or kicking was generally dark when contrasted with undisturbed material nearby.

Summarizing the vertical soil profile of lunar materials as we observed it, the following zones can be defined in order of increasing depth:

(1) A very thin (less than one-eighth in.) light-gray to tannish-gray dusty zone

(2) A thin (about one-fourth in.) dark-gray caked zone that will crack 5 to 6 in. away from where force is applied

(3) A 2- to 6-in.-thick zone of soft, dark-gray to cocoa-gray, slightly cohesive sandy-to-silty material that will hold slopes of probably 70°

(4) A zone that is gradational with zone 3 and is similar to zone 3 except for a marked increase in firmness and resistance to penetration

Zone 3 is the only zone with a thickness that clearly varies with topography, being thickest on the rims of small craters. There was, however, no noticeable thickening of zone 3 on the rim of the large crater just east of the LM. The distribution of rock fragments was not noticeably different among the soil zones, except for an apparently lower concentration in zones 1 to 3 on the rims of the small craters.

Sampling Operations

Contingency Sample

The contingency sample was obtained by taking several partial scoops of soil from zones 1, 2, and 3 in the area in front of quad IV of the LM. There was noticeable difficulty in scooping deeper than about 2 or 3 in. because of an increase in soil firmness. An effort was made to include several small rock fragments in the sample. All the sample was taken from an area that did not appear to have been disturbed by the descent engine gases.

Bulk Sample

The bulk sample was obtained in front of quad IV and the +Y landing gear of the LM. This sample consists largely of scoops of soil from zones 1, 2, and 3 to a depth of about 3 in. and of randomly selected rock fragments from the surface. In some instances it was not possible to scoop more than 1 or 2 in. below the surface; however, there appeared to be no hard layer at that depth.

Documented Sample

The documented sample was taken very rapidly, and no attempt was made to actually document the samples by voice or photography. However, an effort was made to select as wide a variety of samples as possible. The first few rocks were collected from the surface and subsurface 15 to 20 ft north of the LM. In an attempt to avoid the effects of the descent engine gases, the rest of the rocks were collected in the area between the LM and the double elongate crater to the southwest of the LM and in the area between the LM and the Early Apollo Scientific Experiments Package site to the south of the LM.

Core-Tube Sampler

The core-tube samples were taken approximately 10 ft apart about 20 ft northwest of the LM. This area had been generally disturbed by previous operations; however, the samples were not taken directly in footprints.

Photographic Operations

The 70-mm Hasselblad photographs generally were taken using the recommended exposure settings. In some instances on the lunar surface, however, we interpolated between these settings. When exposure conditions were marginal or when a large amount of film remained, we used a range of exposures. Focus settings were selected to give as good a focus as possible at the object of interest, rather than using the preselected focus detents.

Physiological Effects

Distances on the lunar surface are deceiving. A large boulder field located north of the LM did not appear to be too far away when viewed from the LM cockpit. However, on the surface we did not come close to this field, although we traversed about 100 ft toward it. The flag, the television camera, and the experiments, although deployed a reasonable distance away from the LM and deployed according to plan, appeared to be immediately outside the window when viewed from the LM cockpit. Because distance judgment is related to the accuracy of size estimation, it is evident that these skills may require refinement in the lunar environment.

The lunar gravity field also has differing effects on Earth-learned skills. Although the gravitational pull on the Moon is known to be one-sixth of the gravitational pull on the Earth, objects seem to weigh approximately one-tenth of their Earth weight. The mass of an object makes the object easy to handle in the reduced lunar atmosphere and gravitational field. Once moving, objects continue moving, although their movements appear to be significantly slower in the lunar environment. However, personal maneuvering on the lunar surface gives the impression that the gravitational pull is similar to Earth gravity; that is, suited mobility on the Moon does not seem to be much different from unsuited mobility on the Earth.

Finally, the absence of any natural vertical features, coupled with the poor definition of the horizon and the weak gravity indication at the feet of an observer, causes difficulty in identification of level areas when looking down at the surface. The ability to discern level areas is further complicated by the fact that, when a lunar observer is wearing a spacesuit, the center of mass of the lunar observer is higher and farther back than the normal center of mass of an observer on Earth.

Walking in the up-Sun direction posed no problem, although the light was very bright with the Sun shining directly into the visor. While walking in the down-Sun direction, most objects were visible, but the contrast was washed out. Varying shapes, sizes, and glints were more easily identified in the cross-Sun directions.

Although Tranquility Base is a relatively smooth area, the operation of a mobile vehicle in this area would require careful planning. Steep slopes, deep holes, and ridges cover the surface. Special consideration must be given to these characteristics when designing a mobile vehicle that will operate in such an environment.

Solar Corona Observations

During translunar flight, we observed that the solar corona had no noticeable structure or variation in color. The change in illuminosity with radius was similar to that of zodiacal light, especially of long-time-exposure photographs of zodiacal light. Because illuminosity decreased evenly with radius, any estimate of the coronal diameter would depend on how closely one were looking at the illumination; if one were looking very closely, the corona might have a diameter of two, or maybe three, lunar diameters.

3. Geologic Setting of the Lunar Samples Returned by the Apollo 11 Mission

E. M. Shoemaker, N. G. Bailey, R. M. Batson, D. H. Dahlem, T. H. Foss, M. J. Grolier, E. N. Goddard, M. H. Hait, H. E. Holt, K. B. Larson, J. J. Rennilson, G. G. Schaber, D. L. Schleicher, H. H. Schmitt, R. L. Sutton, G. A. Swann, A. C. Waters, and M. N. West

The most urgent tasks of the Apollo Lunar Geology Experiment team have been to obtain information needed for planning the analysis program for the returned lunar samples and to make these data available to other investigators who will study the samples. Consequently, the entire initial effort of the team has been devoted to these tasks. Unusual difficulties have been encountered in this work because of the circumstances of the Apollo 11 mission. For example, time was not available for the astronauts to document, by planned procedures, the localities from which specimens were collected. A critical effort of the experiment team, therefore, has been an attempt to discover the specimen localities by using photographs taken for other purposes. At the time this report was prepared, the attempt had been only modestly successful.

In general, because of the time limitations during the extravehicular activity (EVA), photographs taken on the lunar surface are not well suited for photogrammetric analysis. Data

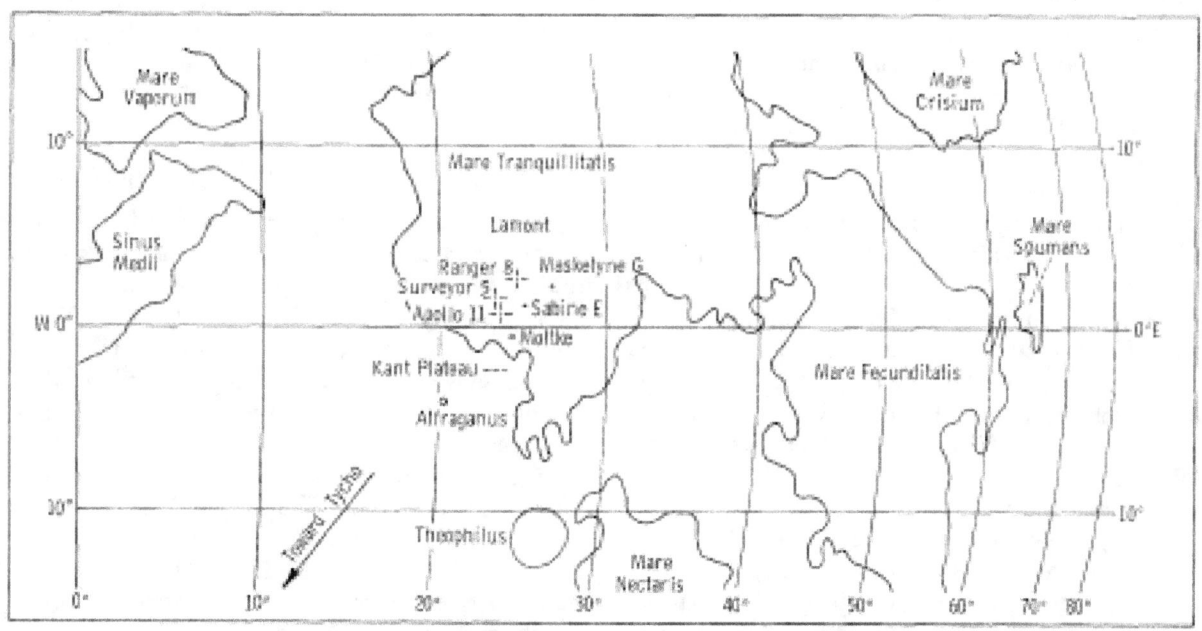

FIGURE 3-1. — Index map of the eastern part of the lunar equatorial belt showing the maria distribution and the location of the Apollo 11 LM and Surveyor 5 landing sites and the Ranger 8 impact crater.

concerning the positions, sizes, and orientation of features on the lunar surface around the lunar module (LM) are being extracted from the photographs by methods that are necessarily laborious. With further work, it is expected that a fairly detailed picture of the topography and geology of the landing site will be recovered and that the original positions on the lunar surface of additional returned specimens will be located. When this has been accomplished, a full interpretive report on the geology will be prepared.

Regional Geologic Setting

The Apollo 11 LM landed approximately 20 km south-southwest of the crater Sabine D in the southwestern part of Mare Tranquillitatis (fig. 3-1). The landing site is 41.5 km north-northeast of the western promontory of the Kant Plateau (ref. 3-1), which is the nearest highland region. The Surveyor 5 spacecraft is approximately 25 km north-northwest of the Apollo 11 landing site, and the impact crater formed by Ranger 8 is 68 km northeast of the landing site.

Mare Tranquillitatis is a "blue" mare (refs. 3-2 and 3-3) of irregular form. The following characteristics suggest that the mare material is relatively thin:

(1) An unusual ridge ring named Lamont, which occurs in the southwest part of the mare, may be localized over the shallowly buried rim of a pre-mare crater.

(2) No large positive gravity anomaly, such as those occurring over the deep mare-filled circular basins, is associated with Mare Tranquillitatis (ref. 3-4).

The southern part of Mare Tranquillitatis is crossed by relatively faint, but distinct, rays trending north-northwest (fig. 3-2) and by prominent secondary craters that are associated with the crater Theophilus, which is located 320 km southeast of the LM landing site. Approximately 15 km west of the landing site is a fairly prominent ray that trends north-northeast. The crater with which this ray is associated is not definitely known; the ray may be related to the crater Alfraganus, 160 km southwest of the landing site, or to Tycho, approximately 1500 km southwest of the landing site. Neither the ray that trends north-northeast nor any of the rays that trend north-northwest cross the landing site; these rays are sufficiently close, however, so that material from Theophilus, Alfraganus, or Tycho is possibly found in the vicinity of the LM landing site. Other distant craters may also be the source of fragments which lie near the LM landing site. In particular, the crater Moltke, which is 40 km southeast of the landing site, may be a source of fragments. Potential distant sources of fragments are in the highlands and in the maria.

On the basis of albedo and crater density, three geologic units can be distinguished in the mare material in the vicinity of the landing site. The LM landed on the most densely cratered unit of these three geologic units. These units may correspond to lava flows of different ages; if this is true, the unit at the landing site is probably the oldest.

A hill of terra material, located 52 km east-southeast of the landing site, protrudes above the mare surface. The occurrence of this hill suggests that the mare material in this region is very thin, perhaps no more than a few hundred meters thick. Craters, such as Sabine D and Sabine E (fig. 3-2), with a diameter greater than 1 km, may have been excavated partly in pre-mare rocks; and pre-mare rock fragments that have been ejected from these craters may occur in the vicinity of the LM landing site.

Geology of the Apollo 11 LM Landing Site

The LM landed approximately 400 m west of a sharp-rimmed ray crater, approximately 180 m in diameter and 30 m deep (fig. 3-3), which had been informally named West Crater. West Crater is surrounded by a blocky ejecta apron that extends almost symmetrically outward approximately 250 m from the rim crest. Blocks as large as 5 m wide occur on the rim and in the interior of the crater. Rays of blocky ejecta with many fragments from ½ to 2 m in width extend beyond the ejecta apron west of the landing point (fig. 3-4). The LM landed in a region between these rays that is relatively free of extremely coarse blocks.

At the landing site, the lunar surface consists of unsorted fragmental debris, which ranges in size from particles that are too fine to be re-

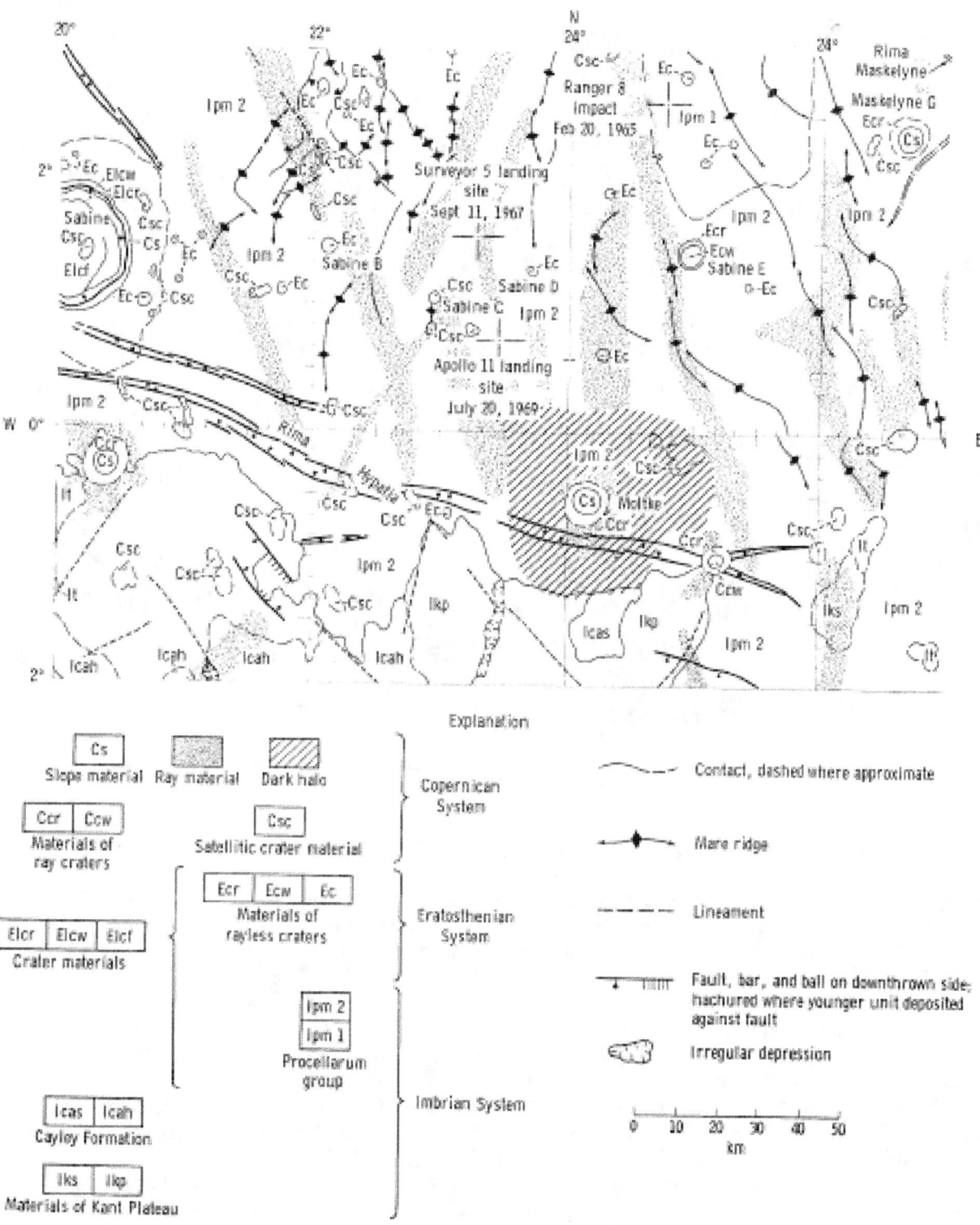

FIGURE 3-2.—Regional geologic map of the region around the LM landing site.

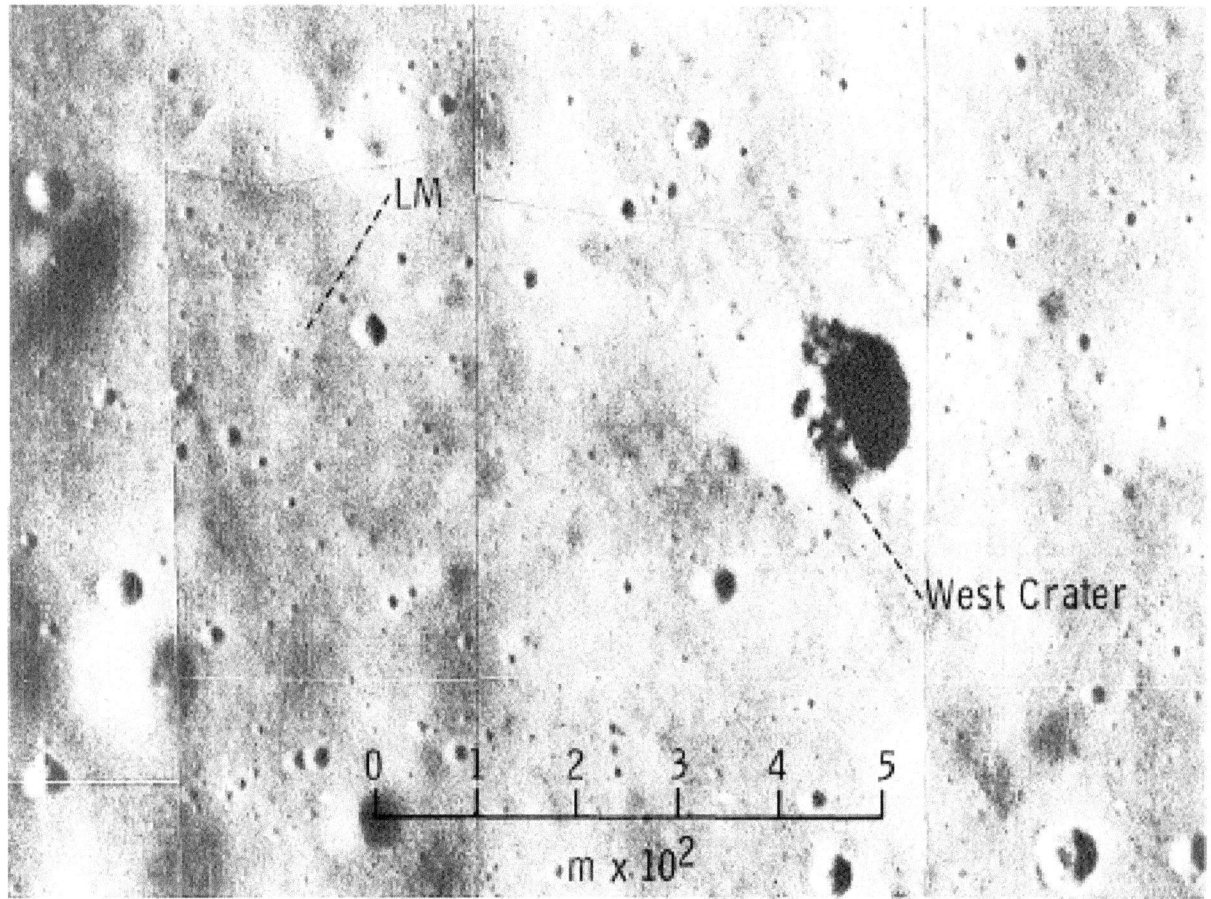

FIGURE 3-3.— Mosaic of Lunar Orbiter 5 photographs showing the location of the LM landing site and of West Crater.

solved by the naked eye to blocks 0.8 m wide. This fragmental debris forms a layer, the lunar regolith, which is porous and weakly coherent at the surface. The regolith grades downward into similar, but more densely packed, material. The bulk of the regolith consists of fine particles, but many rock fragments were encountered on the surface and in the subsurface.

The surface of the regolith is pockmarked with small craters ranging in diameter from only a few centimeters to several tens of meters. Immediately southwest of the LM landing site is a double crater (15 m long, 8 m wide, and 1 m deep) with a subdued raised rim (fig. 3-5). Approximately 60 m east of the LM landing site is a steep-walled, but shallow, crater with a raised rim. This crater, which is 33 m in diameter and 4 m deep, was visited by Astronaut Neil A. Armstrong near the end of the EVA (figs. 3-3 and 3-6).

Many of the small craters have low, but distinct, raised rims; some rims are sharply formed, but most rims are subdued. Other craters are shallow and rimless, or nearly rimless. The small rimless craters are commonly merged together to form irregular shallow depressions. Both the craters and the irregular depressions are distributed without apparent alinement or pattern. Small craters are scattered irregularly on the rims, walls, and floors of larger craters (fig. 3-5).

All the craters in the immediate vicinity of the LM landing site have rims, walls, and floors composed of fine-grained material. Scattered, coarser fragments occur in about the same abundance in these craters as occur on the inter-

FIGURE 3-4. — Hasselblad photograph AS11-37-5468 taken from the LM window and showing a ray of coarse blocks north of the LM landing site.

FIGURE 3-5. — Mosaic of Hasselblad photographs taken from the LM window showing the double crater located southwest of the LM landing site.

FIGURE 3-6. — Mosaic of Hasselblad photographs showing the 33-m-diameter crater east of the LM landing site.

crater areas. These craters are approximately 1 m or less in depth; they have evidently been excavated entirely in the regolith.

In the 33-m-diameter crater east of the LM landing site, the crater walls and rim have the same texture as the regolith has elsewhere; however, a pile of blocks occurs on the floor of the crater (fig. 3-6). The crater floor probably lies close to the base of the regolith. Several craters of about the same size as the 33-m-diameter crater (with steep walls and shallow, flat floors or floors with central humps) occur in the region around the landing site (fig. 3-3). Judging from the depths of these craters, the thickness of the regolith is estimated to range from 3 to 6 m.

An unexpected discovery made by Astronaut Armstrong was the presence of blebs of material with specular surfaces. These blebs of material partially covered 2- to 10-cm-diameter areas in the bottom of six or eight 1-m-diameter raised-rim craters. Astronaut Armstrong observed these apparently glassy blebs, which resembled drops of solder (fig. 3-7), only in craters. The form of the blebs suggests they had been formed by the splashing of molten material traveling at low velocity. The distribution of the blebs suggests that they are natural features on the lunar surface; however, the possibility exists that the blebs are artifacts that were produced by the landing of the LM.

In addition to craters, the surface of the regolith is marked by small, shallow troughs (fig.

FIGURE 3-7.— Closeup stereophotograph AS11-45-6704A showing glassy blebs on the lunar surface.

3–11). By using photographs taken 19 m southeast and 12 m north of the center of the LM landing site, a preliminary study was made of the troughs. Most of the troughs are a fraction of a centimeter to a centimeter deep, approximately ¾ to 3 cm wide, and 3 to 50 cm long; three of these troughs that were observed are 2 to 3 m long. The troughs are located 3 to 5 cm apart in areas in which they are prominent. One set of troughs trends northwest; another set (which is comparable to the northwest-trending troughs in abundance, but are more dispersed in orientation) trends northeast to north-northeast. A few troughs were observed that trend in other directions. Troughs of similar appearance were noted in many other photographs, but the orientations of these other troughs have not yet been determined.

Coarse fragments are scattered in the vicinity

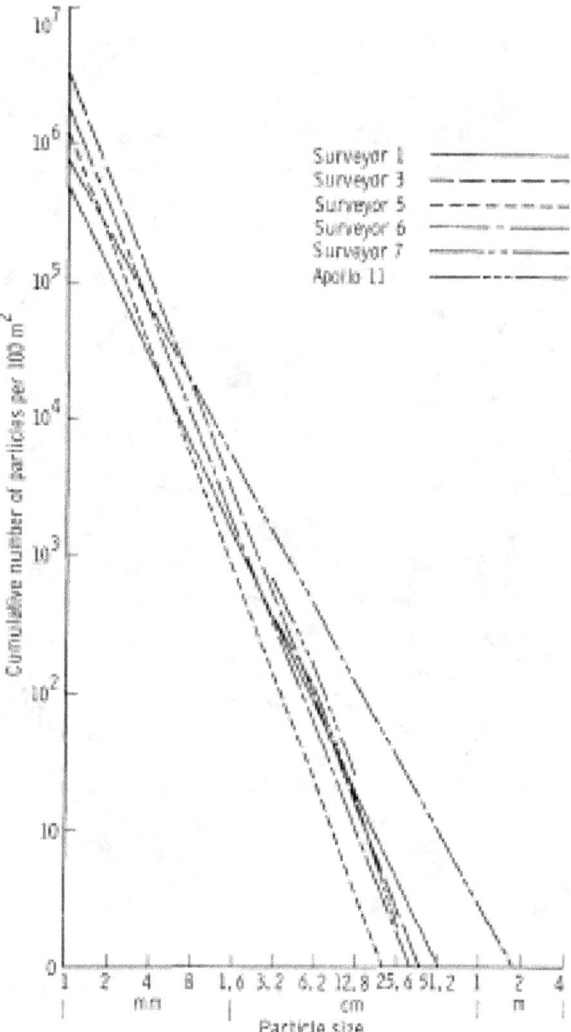

FIGURE 3-9. — Cumulative size-frequency distribution of rock fragments coarser than 3.2 cm at the LM landing site, compared with size and frequency of rock fragments at Surveyor landing sites.

of the LM landing site in approximately the same or somewhat greater abundance than is found at the Surveyor 1 landing site (figs. 3–8 to 3–10) (ref. 3–5). These coarse fragments are distinctly more abundant in these two sites than they are at other Surveyor landing sites on the maria, including the Surveyor 5 landing site, which is northwest of the LM landing site. Similar to the Apollo 11 LM, Surveyor 1 landed near a fresh blocky rim crater but beyond the apron of coarse blocky ejecta. It may be inferred that many rock fragments in the immediate vicinity

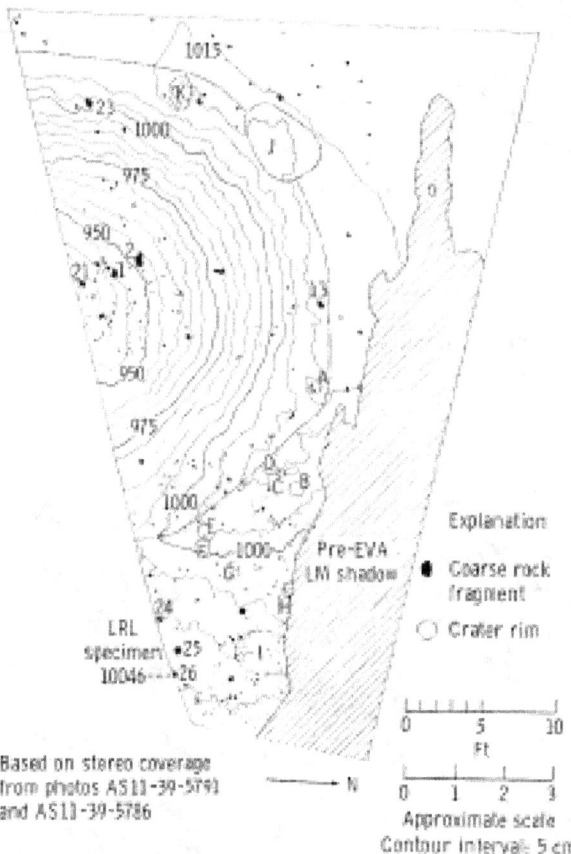

FIGURE 3-8. — Topographic map of part of the lunar surface visible from both windows of the LM showing the distribution of rock fragments.

FIGURE 3-10. — Hasselblad photograph AS11-40-5851 of the region south of the LM landing site from which selected rocks were collected by Astronaut Armstrong.

FIGURE 3-11. — Hasselblad photograph AS11-40-5913 showing part of the lunar surface marked by linear troughs. →

of the spacecraft, at both the Surveyor 1 and the Apollo 11 landing sites, were derived from the nearby blocky rim crater. Fragments derived from West Crater may have come from depths as great as 30 m beneath the mare surface.

The fine-grained matrix of the regolith consists chiefly of microscopic particles. The regolith is weak and easily trenched to depths of several centimeters. Surface material was easily dislodged when kicked. When the flagpole for the U.S. flag and the core tubes were pressed into the surface, they penetrated with ease to a depth of 10 to 12 cm. At that depth, the regolith was not sufficiently strong, however, to hold the core tubes upright; a hammer was needed to drive them to depths of 15 to 20 cm. At several places, rocks were encountered in the subsurface by the tubes, rods, and scoop that were pressed into the subsurface.

The astronauts' boots left prints approximately 3 mm to 3 cm deep in the fine-grained regolith material (fig. 3-12). Smooth molds of the boot treads were preserved in the bootprints, and angles of 70° were maintained in the walls of the bootprints. The fine-grained surficial material tended to break into slabs, cracking as far as 12 to 15 cm from the edges of the footprints.

The finest fraction of the regolith adhered weakly to boots, gloves, spacesuits, handtools, and rocks on the lunar surface. On repeated contact, the coating on the boots thickened until boot color was completely obscured. When the fine particles of the regolith were brushed off, a stain was left on the spacesuits.

In places where fine-grained material was kicked by the astronauts, the freshly exposed material was conspicuously darker than the undisturbed surface. As at the Surveyor landing sites, the subsurface material probably lies at depths no greater than a millimeter from the surface. The existence of a thin surface layer of lighter colored material at widely scattered localities indicates that some widespread process of surface-material alteration is taking place on the Moon.

Fillets of fine-grained material are banked against the sides of most rock fragments. The fillets were observed at least as far as 70 m from the LM, and most of the fillets are almost certainly natural features of the surface. On sloping surfaces, Astronaut Armstrong observed that the fillets were larger on the uphill sides of rocks than on the downhill sides of rocks. The sides of rocks are ballistic traps, and the fillets have probably been formed by the trapping of low-velocity secondary particles. Asymmetric development of fillets around rocks on slopes may be partly caused by preferential downhill transport of material by ballistic processes and partly caused by downhill creep or flow of the fine-grained material.

Disturbances of the Lunar Surface Produced by Landing of the LM

During the landing of the LM, the lunar surface was moderately disturbed by the LM descent-propulsion-system engine blast and by contact of the landing gear with the surface. After the LM had descended to within 30 m of the surface, a sheet of fine dust was observed to move radially away from the center of the rocket plume. Particles composing this sheet, in general, traveled close to the surface and were deflected by the larger rocks.

The descent propulsion system continued to fire a short time after landing; thus, the area beneath the rocket engine is darkened and has a strongly swept appearance. No detectable crater was formed beneath the rocket engine, but a thin, coherent crust appears to have developed.

FIGURE 3-12. — Hasselblad photograph AS11-40-5877 showing an astronaut's bootprint in the lunar surface.

FIGURE 3-13. — Hasselblad photograph AS11-40-5921 showing the lunar surface beneath the LM descent propulsion system.

FIGURE 3-14. — Hasselblad photograph AS11-40-5918 showing the swept ground adjacent to the +Y footpad of the LM.

The crust is locally offset or broken (fig. 3-13). The upper surface of the crust is marked by a large number of fine grooves and ridges that are alined radially with respect to a surface point lying approximately beneath the center of the rocket engine. The radial pattern of grooves and ridges extends at least several meters past the +Y footpad to the north of the LM (fig. 3-14), in the direction from which the LM approached immediately prior to landing. In other directions, the swept ground does not appear to extend much past the footpads.

Probes that extended below the footpads dug only a short distance into the surface and were dragged to the south during the landing. The footpads skidded along the surface (fig. 3-14) and dug in no more than a few centimeters at the leading edge. A small mound of fine-grained material piled up at the leading edge of each footpad.

Surface Traverse and Sampling Activities of the Astronauts

The surface traverse and sampling activities of the astronauts have been reconstructed from clues provided by the voice transcript, from review and analysis of the lunar television pictures, from analysis of the 16-mm pictures taken with a time-sequence camera mounted in the LM cockpit, and from detailed study of photographs taken with Hasselblad cameras before, during, and after the EVA.

The camera stations for Hasselblad survey panoramas taken on the EVA were located by photographic triangulation from mosaics. Horizontal angles between the LM footpads were measured on the photographs as a function of the known field of view, and the angles were drawn on tracing paper. The paper was then manipulated over a scale drawing of the LM until the lines intersected the appropriate pads at the proper place. Once the panorama locations had been determined, azimuths were measured from two or more panoramas to conspicuous features on the surface, and the positions of the features were plotted by triangulation to produce the map of figure 3-15. Individual photographs were located, and their orientations were measured by similar methods, using for control both the LM footpads and other features on the lunar surface that had been located by triangulation from the panoramas.

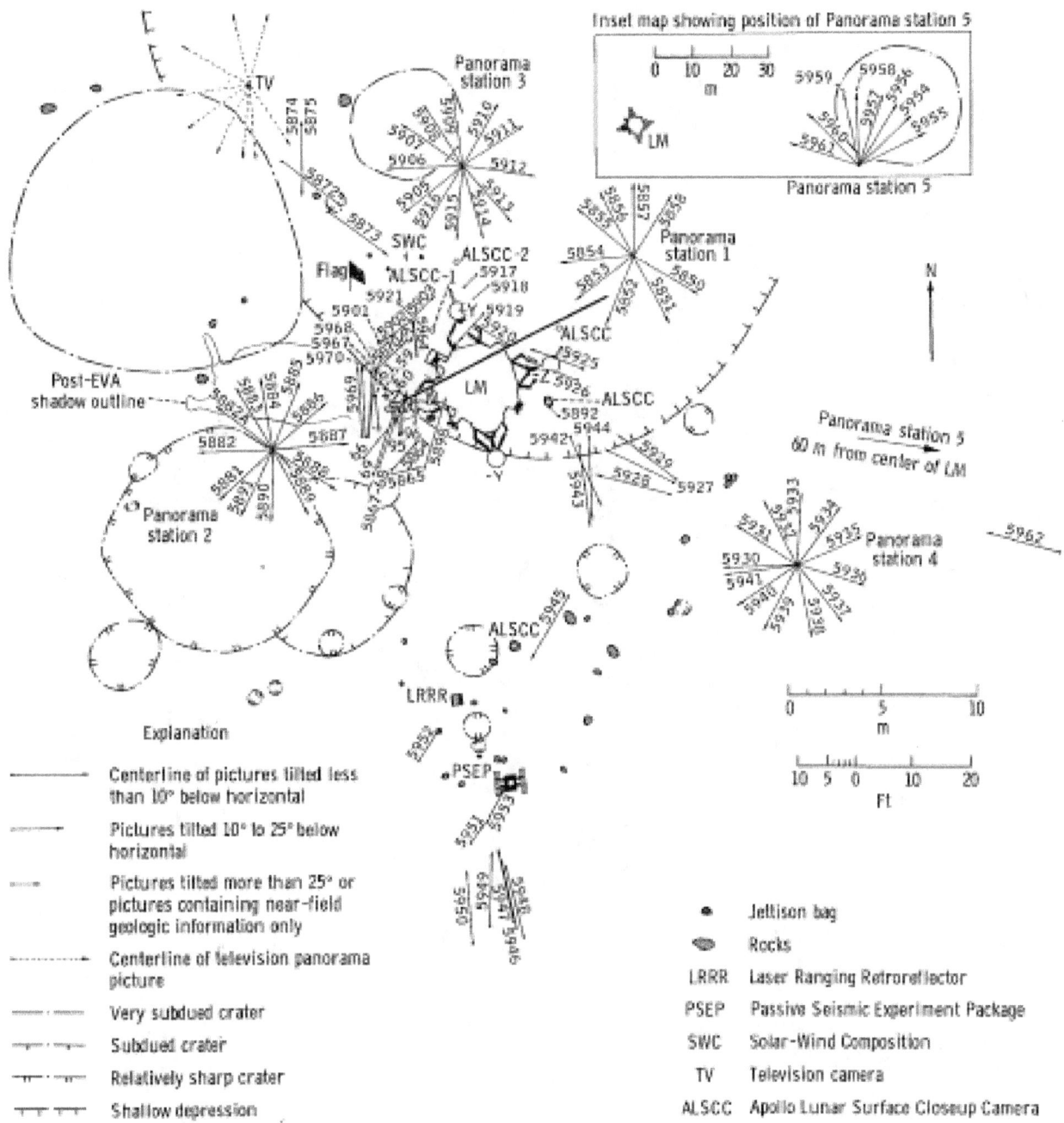

FIGURE 3-15.—Preliminary map of EVA photographs and television pictures taken at the landing site.

Because the graphical method by which these data have been obtained is fairly crude, azimuths shown for individual frames may have errors of 3° or more. Positions of most of the camera stations are probably within a 1.5-m circle centered at the point shown. The determinations are sufficiently accurate, however, to provide a useful control net for an overall view of the astronauts' traverse and a starting point for more rigorous analytical photogrammetric measurements.

A traverse map has been prepared (fig. 3-16), using for control the object locations obtained by graphic solution. The plot of the surface traverse of the crew was obtained primarily from

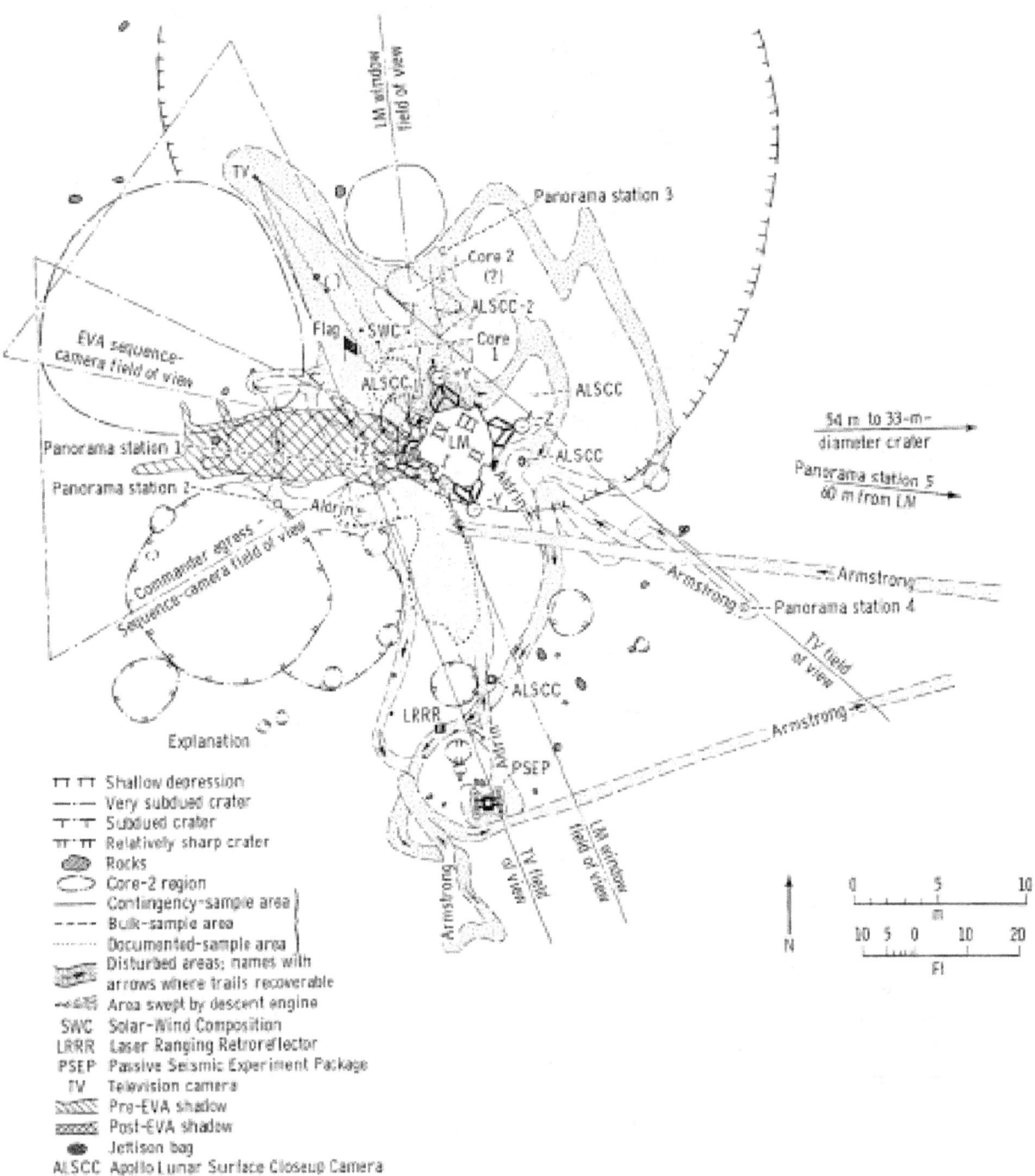

FIGURE 3-16. — Preliminary traverse map of the landing site.

study of the Hasselblad photographs. Regions disturbed by walking are easily detected because of their lower albedo. A rough estimate of the minimum total distance traveled by Astronauts Armstrong and Aldrin combined is 750 to 1000 m. The most distant single traverse was made by Astronaut Armstrong to the 33-m-diameter crater east of the LM.

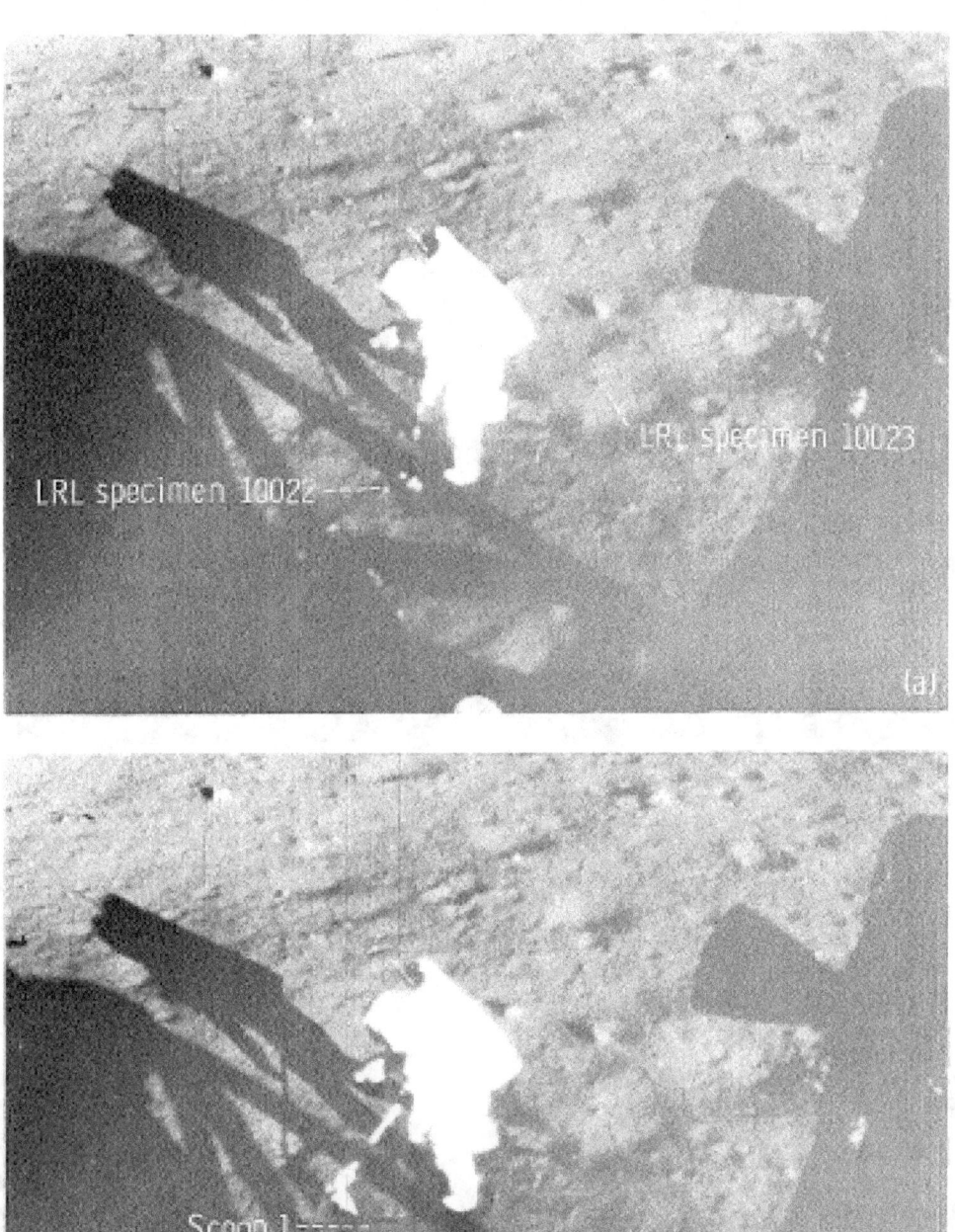

FIGURE 3-17.—Enlargement of 16-mm sequence-camera photographs (magazine K) showing Astronaut Armstrong immediately prior to and immediately after scooping up two rock samples in the first of two contingency sample scoops. (a) Immediately prior to scooping. The larger rock (6 cm in width) has been identified as LRL specimen 10022 (fig. 3-23). (b) Immediately after scooping. Note the scoop mark where the two rocks had been lying on the lunar surface.

FIGURE 3-18.— Enlargement of 16-mm sequence-camera photographs (magazine K) showing Astronaut Armstrong immediately prior to and immediately after taking the second contingency sample scoop. (a) Immediately prior to second scooping. Note LRL specimen 10023 below the scoop. (b) Immediately after second scooping. Note LRL specimen 10023 is missing, and a shallow scoop mark is visible.

Sampling Log

Contingency Sample

The contingency sample was collected in full view of the sequence camera immediately northwest of the LM, and the collection time was approximately 3 min and 35 sec. The sample bag was filled with two scoops of sample having a total weight of approximately 1.4 kg. From study of the sequence film data (figs. 3-17 and 3-18), the regions that were scooped have been accurately located on a pre-EVA LM window photograph. Both scoops included small surface rock fragments that were visible from the LM windows prior to sampling (fig. 3-19).

Bulk Sample

Astronaut Armstrong required 14 min to collect approximately 16 kg of bulk sample, from the beginning of sampling to sealing of the Sample Return Container (SRC). Of that time, 5 min were spent in sealing the SRC. Astronaut Armstrong went out of the television field of view three times during bulk sampling; he traversed twice to the left of the field of view for a total of 1 min and 11 sec and once to the right of the field of view for 35 sec. At least 17 or 18 scoop motions were made in full view of the television camera, and at least five scoop motions were made within the field of view of the sequence camera (fig. 3-20). The total number of scoops was 22 or 23. Nine trips back to the Modularized Equipment Stowage Assembly (MESA) were made to empty the scoop. An average of 2½ scoop motions was required to fill the scoop.

Documented Samples

A total of 5 min and 50 sec was spent by Astronaut Aldrin in obtaining the two core-tube samples, both of which were taken in the vicinity of the Solar Wind Composition Experiment. The driving of the first core was documented by the television camera and by two individual Hasselblad photographs. The second core-tube location

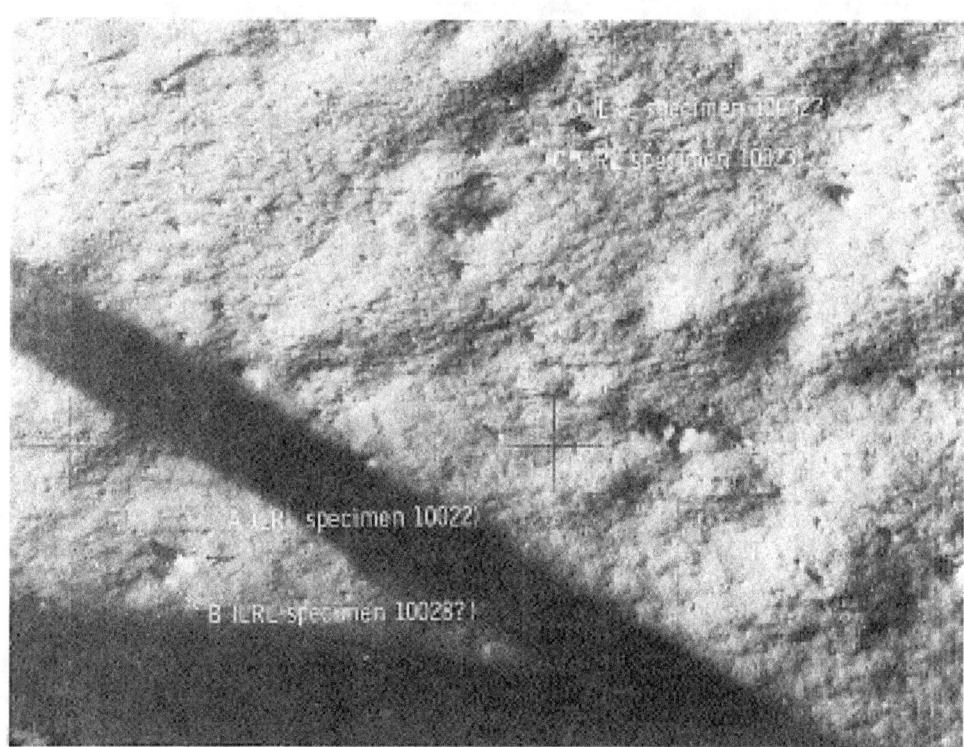

FIGURE 3-19.— Hasselblad photograph AS11-39-5777 taken from the LM window showing rocks A, B, C, and D, which were collected in the contingency sample scoops.

(in the vicinity of the Solar Wind Composition Experiment) was confirmed only by Astronaut Aldrin's verbal statement to that effect (ground-elapsed time 04:15:19:19).

Only 3 min and 35 sec (the same time as for the contingency sampling) was available for collection of approximately 20 selected, but unphotographed, grab samples (approximately 7 kg) by Astronaut Armstrong in the final minutes of the EVA. Collection of these specimens was made a distance up to 10 to 15 m away from the LM in the region south of the +Z strut near the east rim of the large double crater (fig. 3-16). Astronaut Armstrong was outside the television field of view (to the west) 25 percent of the time during this activity.

Location of Orientation of Rock Specimens

A preliminary search has been made for individual rock fragments collected in the contingency, bulk, and documented samples. Two rocks (Lunar Receiving Laboratory (LRL) specimens 10022 and 10023) in the contingency sample and one large rock in the bulk sample (LRL specimen 10046) have so far been located and oriented. Two other rocks in the contingency sample (LRL specimens 10028 and 10032) have been tentatively identified. Many other rocks have been determined to be missing

FIGURE 3-20. — Hasselblad photograph AS11-39-5802 shows scoop sites and scoop movements (indicated by arrows) of five of the approximately 22 bulk sample scoops, as determined from the sequence-camera photographs. Numbers in parentheses indicate the nine trips that were made taking the samples back to the MESA area of the LM to empty the scoop. Numbers without parentheses represent the order in which scoops were taken in the view of only the sequence camera. Most of the trenches left by scooping were destroyed later. Scoops 2, 3, and 5 are within the view of the camera, and the trenches made by these scoops are still visible. Rocks were sampled in scoops 4 and 5.

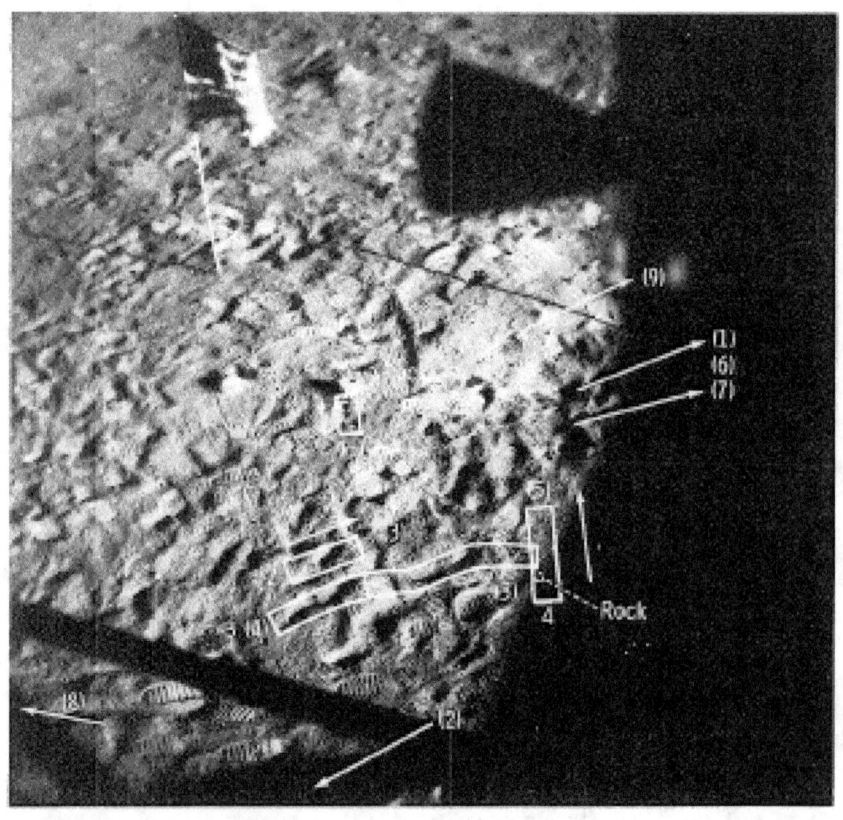

and possibly sampled; but they are not yet correlated with specimens returned to the LRL.

Hasselblad photographs taken from the LM windows before and after EVA, EVA sequence camera photographs, and the first panorama taken with a Hasselblad camera during the EVA have been used to locate scoop sites and rock fragments for both the contingency and the bulk samples. The panorama taken by Astronaut Armstrong soon after he stepped onto the lunar surface provided a near 360° view of the undisturbed surface. This surface perspective, complemented by Hasselblad photographs taken from the LM window before EVA, provided the information used for identification of individual rocks.

Contingency Sample

Both contingency sample scoops were taken within view of the sequence camera, and the lunar surface regions scooped were located on pre-EVA Hasselblad photograph AS11-39-5774 by reference to small craters and rocks in the vicinity. (Hasselblad photograph numbers refer to photographs on file with the NASA Manned Spacecraft Center Photographic Laboratory.) The locations of two exposed rocks sampled during scoop 1 and another exposed rock sampled during scoop 2 were verified by photographs showing their presence before the scoop motion and their absence afterward (figs. 3-17 and 3-18). A fourth exposed rock, too small to be resolved on the 16-mm sequence-camera film, has been tentatively identified in the LM window Hasselblad photographs as having been collected in scoop 2. Distinct scoop marks were visible at these locations immediately after sampling, but were destroyed by later EVA activities in the area. Figure 3-19 shows the four exposed rocks obtained during contingency sampling, as viewed from the LM window. Figure 3-21 and figure 3-22, which were taken as part of the first panorama, also show the four indicated contingency rocks.

The next step in identification of the contingency rock samples was to obtain an approximate size of each specimen on the lunar surface by comparison with bootprints in close proximity. The medium lunar surface boot is 35.5 cm long and 16.0 cm wide at its widest point. In

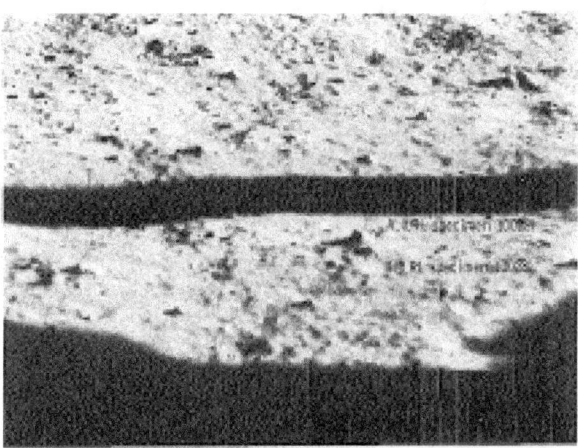

FIGURE 3-21. — Enlargement from Hasselblad photograph AS11-40-5837 showing rocks A and B, which were sampled in the first contingency sample scoop. Rock A has tentatively been identified as LRL specimen 10022. (Compare with fig. 3-33.) Rock B has not been satisfactorily identified, but may be LRL specimen 10028.

figure 3-19, rock A is approximately 6 cm, rock B is approximately 2 cm, and rock C is approximately 5.5 cm, each in its longest exposed dimension.

Photographs taken in the LRL show that four fragments coarser than 4 cm were recovered from the contingency sample. The largest fragment (LRL specimen 10021) is approximately 9 cm long, and two fragments (LRL specimens 10022 and 10023) are 6 to 6⅜ cm long. One fragment (LRL specimen 10024) is 5 cm long.

Figure 3-19 shows that rock A is an angular fragment with a characteristic pointed end. LRL specimen 10022 is a vesicular, crystalline rock with the proper shape and size to match rock A. The two remaining rocks that are at least 6 cm long do not have a pointed end of the proper shape. A preliminary orientation of LRL specimen 10022 is shown in figure 3-23.

Rock B (fig. 3-19) is approximately 2 cm in its largest dimension and has a distinct groove in the upper surface that is visible in photographs taken from the LM window. This rock has been tentatively correlated with LRL specimen 10028, which measures 2.5 by 2 by 1 cm and is the eighth largest fragment in the contingency sample.

Rock C (figs. 3-19 and 3-22), which was col-

FIGURE 3-22. — Enlargement from Hasselblad photograph AS11-40-5857 showing the rocks that were sampled in the second contingency sample scoop. The rocks have been tentatively identified as LRL specimens 10023 and 10032.

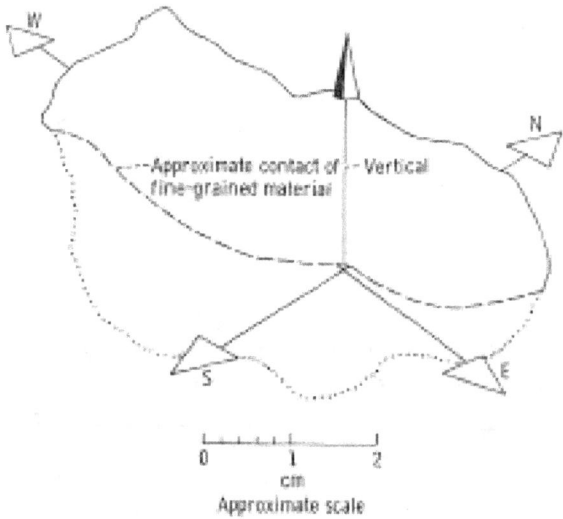

FIGURE 3-23. — Approximate orientation of LRL specimen 10022 in lunar coordinates.

lected in the second contingency sample scoop, has a characteristic deep depression 1.2 to 1.5 cm in diameter on its exposed face. This depression facilitates identification and orientation.

Measurement of the rock from the Hasselblad photographs indicates that the exposed part is approximately 5 cm long. The LRL specimens 10021 and 10023 were studied as possible matches because they were the only two remaining rocks in the contingency sample that were large enough. The LRL specimen 10021 was rejected because it does not contain a sufficiently deep depression of the proper size (1.2 cm) to match that observed on the sampled rock. The LRL specimen 10023, however, contains a deep pit of the appropriate size. An approximate orientation of LRL specimen 10023 with respect to its position on the lunar surface is shown in figure 3-24.

Less than 5 cm from rock C is a small rock 1½ cm wide (rock D on figs. 3-19 and 3-22) that was almost certainly collected in the second contingency sample scoop but is too small to be clearly resolved in the 16-mm sequence-camera photographs. On the basis of size and shape, this rock has been tentatively correlated with LRL specimen 10032, which is the 12th largest fragment in the contingency sample.

Bulk Sample

During collection of the bulk sample, five separate scoop marks were made within view of the EVA sequence camera. The location of these marks on the lunar surface has been verified (fig. 3-20). At least two small rocks, 1 to 2 cm wide, were included in scoops 4 and 5 (fig. 3-20) but are too poorly resolved to be identified in the photographs. Larger rocks collected in the bulk sample include three or four rocks approximately 10 cm wide, five rocks approximately 5 cm wide, and 15 or more rocks that range from 1 to 5 cm wide. Thus far, only one rock 10 cm wide has been located and oriented.

A subrounded, equidimensional rock was determined to be missing from a point approximately 4 m from the +Z footpad at a 250° azimuth (fig. 3-25). The rock was located by comparison of pre-EVA and post-EVA LM window photographs. (For an example, see fig. 3-26.) A study of figure 3-27, taken after bulk sampling, verified that the rock had been removed by the bulk sampling scoop. A scoop mark is clearly visible in figure 3-27 in the position that the rock had occupied. Measurement

FIGURE 3-24.—Orientation diagram of LRL specimen 10023. (a) The specimen is shown as viewed from the southeast. The rounded surface and the approximate intersection of the rock surface with the lunar surface is shown. Orientation of the specimen in lunar coordinates is shown. At top is LRL photograph S-69-45394. (b) This diagram shows the flat, northwest face of the sample and the approximate intersection of the rock surface with the lunar surface. Orientation of the specimen in lunar coordinates is shown. At top is LRL photograph S-69-45392.

of the specimen by comparison with bootprints shows that it is 10 cm wide. The LRL photographs were searched for the larger rocks collected in the bulk sample. The northeast face of the rock as seen in photograph AS11-40-5853 (fig. 3-28) was readily identified in the LRL photograph S-69-45610 of LRL specimen 10046 (fig. 3-28).

Documented Sample

Several rocks in the documented sampling area south of the +Z footpad have been found to be missing, moved, stepped on, or possibly sampled. As yet, no definite correlations between missing rocks and LRL specimens have been made, but it is possible that several larger specimens will be found after closer study.

The pre-EVA and post-EVA LM window photographs show the southeastern part of the sampling area (figs. 3-29 and 3-30). Missing rocks, labeled E, F, and G in figure 3-29, are clearly visible in photograph AS11-40-5945 taken during the EVA (fig. 3-31). Rock E appears to have been stepped on and broken, while rocks F and G have definitely been removed from their original position. Several small elongate depressions near rocks F and G may be seen in the post-EVA photograph (fig. 3-30).

FIGURE 3-25.—Enlargement from Hasselblad photograph AS11-40-5853 showing undisturbed rocks and LRL specimen 10046.

FIGURE 3-26.—Enlargement from Hasselblad photograph AS11-37-5502 showing the disturbed rocks and changes that occurred during EVA.

FIGURE 3-27. — Hasselblad photograph AS11-40-5887, which was taken from the second panorama-photograph station, showing the region from which LRL specimen 10046 was taken.

These depressions may have been formed by the dragging of the specimens across the surface as they were being collected. Rock F, which is approximately 20 cm across, apparently was not returned in the documented sample container, and it is suspected that this rock corresponds to a large fragment picked up with the sampling tongs and then dropped by Astronaut Armstrong in this vicinity during documented sampling.[1] With more detailed study, there is a good possibility that rock G can be correlated with an LRL specimen.

Rock Fragments

Individual specimens of returned rock have been studied by the Lunar Geology Experiment team by comparing Hasselblad photographs of the rocks in their original positions on the lunar surface and photographs of the rocks taken in the LRL. The purpose of these studies was to determine the local geologic environment of each specimen and the relationship of each specimen to other rocks and materials at the sample locality. From the field geologic relations and the hand specimen characteristics of each rock, an attempt has been made to extract the data that bear on geologic processes that have

[1] Personal communication from Astronaut Armstrong, Aug. 27, 1969.

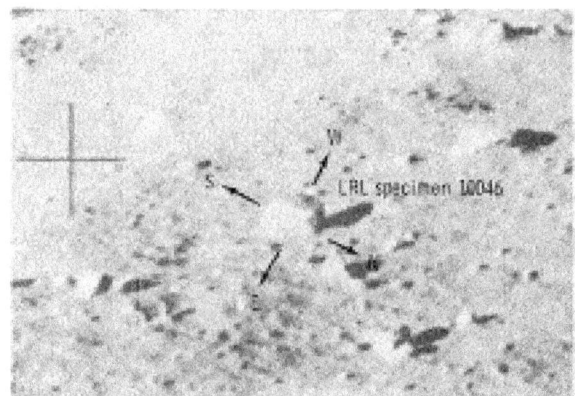

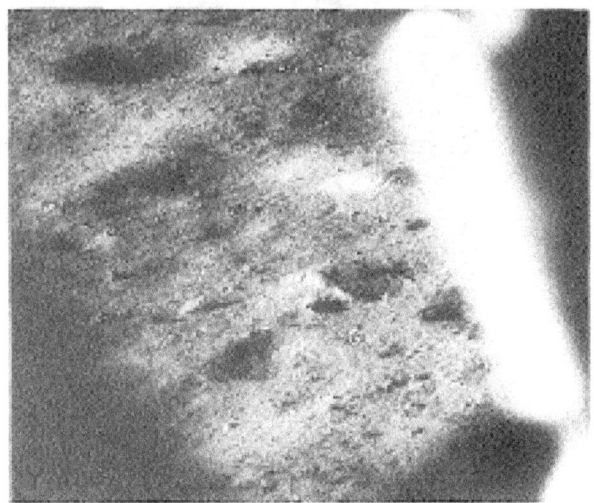

FIGURE 3-29. — Hasselblad photograph AS11-39-5147 taken from the LM window looking south. This photograph shows rocks F and G, which were probably moved during the documented sampling period.

FIGURE 3-28. — Hasselblad photograph AS11-40-5853 and LRL photograph S-69-45610 of LRL specimen 10046 showing the lunar orientation of the specimen.

affected the specimen and to extract data on the geologic history of the specimen.

The data presented have been derived entirely from photographs and have been obtained independently from mineralogical and other investigations conducted in the LRL. Each returned rock specimen is considered to be a geologic entity with a unique history, and the specimen is investigated as a whole, as it was collected on the Moon, rather than analyzed for its components. Of particular interest are the diverse surfaces of each rock, from which much can be learned about the history of the rock after it became a fragment in the lunar regolith.

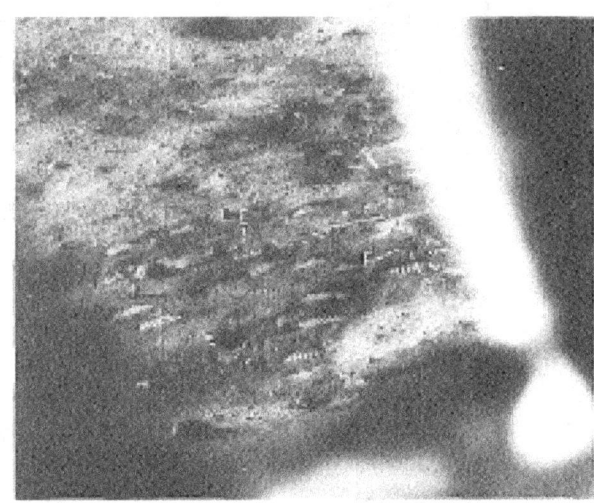

FIGURE 3-30. — Hasselblad photograph AS11-39-5794 taken from the LM window looking south showing the region where rocks F and G are missing.

By studying many rocks from this standpoint, the processes may be elucidated by which the lunar regolith and its constituent fragments were formed.

Rock fragments at the Apollo 11 LM landing site have a wide variety of shapes, and most of the fragments are embedded to varying degrees

FIGURE 3-31.— A section of Hasselblad photograph AS11-40-5945 showing the detail of rocks F and G.

in the fine matrix of the regolith. A majority of the rocks are rounded to subrounded on their upper surfaces, but angular fragments of irregular shape are also abundant. A few rocks are rectangular slabs with a faint platy structure. Many of the rounded rocks, when collected, were found to be flat or of irregular angular shape on the bottom. The exposed part of one unusual rock, which was not collected, was described by Astronaut Armstrong as resembling a distributor cap. When this rock was dislodged with a kick, the sculptured "cap" was found to be the top of a much bigger rock, the buried part of which was larger in lateral dimensions and angular in form.

The rocks sampled are of two basic types: (1) rocks with crystalline textures of fine- to medium-grain size and (2) aggregates of clastic material. The most abundant of the returned crystalline rocks are fine-grained rocks with vesicular or platy textures that resemble terrestrial basalts. During his traverse on the Moon, Astronaut Armstrong called these rocks basalts and commented on the vesicles in some of them. One of the best examples of these basaltlike materials is LRL specimen 10022, which is described in a later section of this report.

Subordinate in number to the basaltlike specimens are medium-grained rocks that are composed dominantly of dark-colored minerals plus a white mineral, which may be plagioclase. When interpreted in terms of terrestrial rocks, these lunar rocks resemble diabases and gabbros (LRL specimen 10047), and feldspathic pyroxenites or peridotites (LRL specimen 10044). The grain size of these specimens is 0.5 to 2 mm. The texture, as revealed under stereoscopic examination of the photographs, consists of large amounts of a euhedral dark-colored mineral, which may be clino-pyroxene, with two cleavages at about right angles and a white cleavable mineral, which may be plagioclase. Some rocks contain rounded dark-colored minerals surrounded by jackets of the plagioclaselike and pyroxenelike minerals. In LRL specimen 10044, moreover, a stratiform banding is visible, although it is difficult to distinguish from exfoliation shells that are present in the same specimen. These features are similar to the textures and banding found in specimens from mafic plutonic complexes on Earth.

Aggregates of weakly to strongly indurated clastic materials, which are composed mostly of rough-surfaced spheroidal to angular grains and of clasts ranging from less than 0.1 to more than 1 cm in diameter, constitute about half of the returned rocks. These rocks appear to have been formed by induration of the lunar regolith material. Microbreccia is an appropriate term for them. Typical specimens, which are described in a following section of this report, are LRL specimens 10023 and 10046. In two of the rocks studied, the material appears to show stratification (at the small scale of the specimen). Many rounded grains are too small for their true character to be identified in photographs. Some specimens appear to be composites, perhaps made by cementation of silt-sized particles; others appear to be rock fragments.

Minute deep pits, ranging from a fraction of a millimeter to approximately 2 mm in diameter, occur on the rounded surfaces of most rocks. These pits were observed by Astronaut Armstrong, who recognized that they had been produced on the surface of the rocks and that they were distinguishable from vesicles. Many of these pits are lined with glass. They are present on two microbreccia specimens (LRL specimens 10023 and 10046) that have been identified in photographs taken on the lunar surface and for which the orientation of the rocks at the time they were collected is known. The pits are found primarily on the upper sides of these specimens; therefore, they clearly have been produced by a process acting on the exposed surface. The pits do not resemble impact craters produced in the laboratory, and their origin has not been determined.

On the rounded surfaces of some rocks, a thin, altered rind has developed. The rind, which has been observed on both the crystalline rocks and the microbreccias, is characteristically approximately ½ to 1 mm thick. On freshly broken faces, the rind is observed as a thin, light-colored zone. The rind appears to be caused in part by the shattering of mineral grains. Halos of light-colored material, which is similar to the rind in appearance, occur around the pits on many specimens.

The rounding of the exposed upper surfaces of many rocks suggests that processes of erosion that lead to gradual rounding are taking place on the lunar surface. Several processes may be involved. On some rounded rock surfaces, individual clasts and grains (of which the rocks are composed) and glassy linings of the pits on the rock surfaces have been left in raised relief by general wearing away or ablation of the surface. Rocks that exhibit this differential erosion most prominently are microbreccia. The ablation may be caused primarily by small-particle bombardment of the surface.

Exfoliation is another major process of rounding. The rounded forms that characterize the top surfaces of some of the rocks collected are clearly caused by the peeling off of closely spaced exfoliation shells. Examples of rocks rounded by exfoliation are LRL specimens 10044 and 10047. The curved shells passing completely around the specimen cross the mineral banding of the stratiform rock (LRL specimen 10044) without deviation and round off the corners of the rock. The distributor-cap form observed by Astronaut Armstrong may have developed by exfoliation and could have been produced by spalling of the free surfaces of the rock as a result of one or more energetic impacts on the top surface.

The following paragraphs provide descriptions of LRL specimens 10022, 10023, 10032, 10046, and 10047 and tentative identification of the regions of the lunar surface from which the samples were selected.

Geologic Specimen Descriptions

LRL Specimen 10022

Geologic environment. LRL specimen 10022 is the third largest fragment collected in the contingency sample. The specimen has been tentatively identified in photographs of the lunar surface, pending confirmation by photogrammetric measurements. The photographic coverage of the specimen locality is good. Photographs AS11-39-5773, AS11-39-5774, and AS11-39-5777, taken from the LM windows, provide stereoscopic coverage of the locality, but only with a short stereographic base. Photograph AS11-40-5857, taken from near the foot of the spacecraft ladder (panorama I), also shows the specimen.

A region approximately 1.1 by 2.4 m surrounding LRL specimen 10022 was studied from the photographs. The specimen is located approximately 2.5 m north of the center of the spacecraft +Z footpad. Parts of the locality are obscured by shadows cast by the +Y landing struts of the spacecraft. The region studied is relatively level, with no pronounced topographic features (fig. 3-32). Areas enclosed by the break-in-slope symbol in figure 3-32 are coalescing, subdued depressions. Located approximately 4 cm northeast of LRL specimen 10022 is a sharp irregular depression approximately 10 cm across, which is the only depression of this type in the region studied. Elongate troughs measuring a fraction of a centimeter deep, ½ to 2 cm wide, and 3 to 20 cm long are prominent features at this locality. While most of the largest troughs trend northeast, another set of troughs trends northwest. Examination of photographs taken during the EVA indicates that similar troughs are present at least as far as 60 m from the LM.

Approximately 280 fragments between ½ and 2 cm, and 11 fragments (including LRL specimen 10022) between 2 and 8 cm wide, are scattered over the region studied. Of the 11 larger fragments, LRL specimen 10022 and one smaller fragment are angular; the other nine

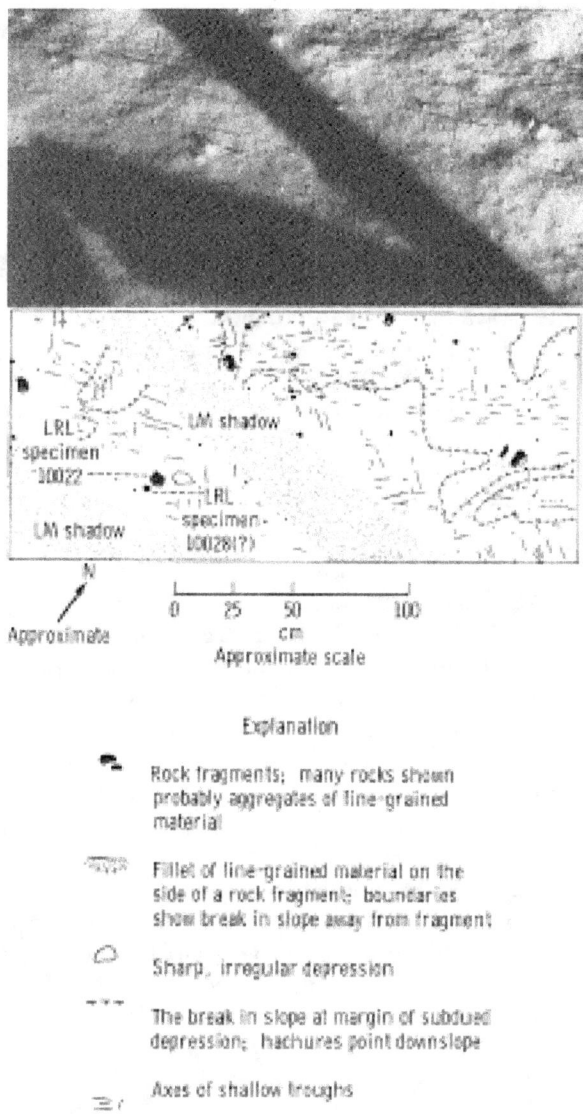

FIGURE 3-32.—Hasselblad photograph AS11-39-5777 and map showing the region around LRL specimens 10022 and 10028 that was studied.

cm wide are buried appears to vary uniformly from almost total burial to no burial.

All the larger fragments have fillets of fine-grained material banked against the sides that are observable in the photographs. The fillets are approximately 1 to 1½ cm high on the four largest fragments and are proportionately smaller on the other seven fragments.

The 11 fragments that measure larger than 2 cm wide have no apparent relationship to surrounding features, except for their own fillets. The clusters of small fragments lying in the west and northwest sectors of the center of the map area may be remnants of low-velocity impact projectiles that formed the craters.

LRL specimen 10022 lay within a level part of the region, approximately 30 cm from the edge of the nearest subdued depression. No apparent relationship exists between the specimen and the subdued depressions. It is conceivable that the sharp irregular depression 4 cm to the northeast of LRL specimen 10022 is a secondary crater from which the specimen skipped. However, the presence of a fillet of fine-grained material against the specimen suggests that it had been in place for an appreciable length of time, probably sufficiently long for any secondary crater that may have been caused by the specimen to be more subdued than the sharp depression by surface processes.

LRL specimen 10022 is roughly tabular in form and was slightly embedded in the regolith, tilted southeast at a low angle to the surface. The exposed part of the specimen is angular, and the unexposed part is rounded (fig. 3-33). The fillet of fine-grained material was approximately 1 cm high and 1½ cm wide, and the fillet covered the southern part of the upper surface of the rock. Little of the rock lay below the general level of the surrounding surface. The fillet-rock contact is marked by a light-colored zone approximately one-third cm wide on the recovered specimen (fig. 3-33).

The fragment nearest LRL specimen 10022, located 2 cm south of LRL specimen 10022, was also collected in the contingency sample and has been tentatively identified as LRL specimen 10028. LRL specimen 10022 is one of two crystalline rocks of significant size in the contingency sample and is dissimilar in appear-

fragments are rounded. The 11 larger fragments appear to be randomly scattered over the region studied. Fragments 1 to 1½ cm wide are concentrated more heavily than average in two areas within subdued depressions located west and northwest of the center of figure 3-32.

Of the 11 fragments larger than 2 cm wide, five rest on the surface, and as many as one-third to one-half of the other six appear to be buried. The extent to which fragments smaller than 2

FIGURE 3-33.— LRL photographs S-69-45209 and S-69-45210 of LRL specimen 10022. (a) LRL photograph S-69-45209. (b) LRL photograph S-69-45210.

ance to LRL specimen 10028 and to all the other fragments in the contingency sample. LRL specimen 10022 is the most angular of the 11 largest fragments in the region studied, which suggests that LRL specimen 10022 may have been in its photographed position on the lunar surface for a shorter period of time than other fragments of comparable size. However, the crystalline nature of the specimen suggests that it may be more resistant to erosion, which may also account for its higher degree of angularity. The more rounded, buried part of the specimen suggests that the underside was at one time the upper surface and that the underside has been exposed to lunar surface erosion processes for a longer time than the present upper side.

Description. LRL specimen 10022 is a smaller (2 by 4 by 6 cm) specimen of fresh-appearing vesicular lava and is similar in vesicularity, texture, and crystallinity to many terrestrial basalts. Smooth-walled vesicles ranging from 0.1 to 4 mm in diameter occupy approximately 15 percent of the rock by volume. Some vesicles are well oriented into vesicle trains; the vesicles flatten and partially merge, near one edge of the specimen, into an incipient zone of platy jointing. Except where several vesicles have merged into vuglike cavities, the majority of the vesicles are either nearly spherical or have been pulled into prolate spheroids that are apparently oriented by flow. The vesicle walls appear to be smooth, but slight modification shows that at least some of them are coated by a botryoidal substance with a vitreous luster.

Although the minerals are extremely fine-grained, the apparently intergranular to diktytaxitic texture typical of basalt lava can be seen in the photographs. The texture appears identical to that of many fresh olivine basalts from the Oregon Cascades. Microlites appear to have been oriented by flow, and the flow lines deviate around many of the vesicles in the same pattern common to terrestrial basalts.

The outside of this specimen shows a higher albedo than the interior. This condition was possibly caused by the opening up of the cleavages, grain boundaries, and tiny incipient cracks in the rock. This whitish rind, which is barely noticeable on the edges of the broken surface in LRL photograph S-69–45380, penetrates the rock to a depth of approximately 1 mm. Pits filled with glassy material and small fractures containing glassy materials occur on the surface of the specimen.

LRL Specimens 10023 and 10032

Geologic environment. LRL specimen 10023 was collected approximately 4 m west of the +Y footpad. LRL specimen 10032 is believed to have been collected within a few centimeters of LRL specimen 10023. A region approximately 2.5 by 3 m and centered on the rock (fig. 3–34) was studied to determine the geologic environment of these specimens. The region is gently rolling with several broad, subdued, and irregular depressions and four well-defined circular

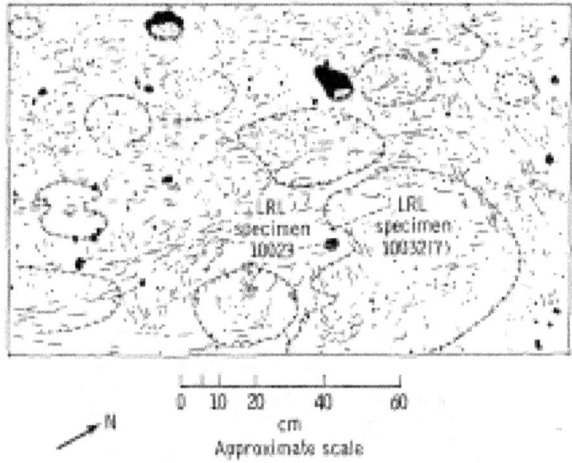

FIGURE 3-34.—Hasselblad photograph AS11-39-5774 and map showing the region around LRL specimens 10023 and 10032 that was studied.

craters. One circular crater has a distinct raised rim around three-quarters of its circumference. LRL specimens 10023 and 10032 are embedded on the western edge of a subdued, irregular depression in a part of the surface that slopes gently to the east.

Two well-defined sets of linear features extend across the region studied. The first set consists of both slightly curved and straight troughs that range in length from approximately 2 cm to approximately 10 to 15 cm and trend in a northeasterly direction. This set is best seen when figure 3-34 is viewed obliquely from the upper right-hand corner.

Within the northeast set of lineations, a north-trending zone approximately 15 to 20 cm has wide straight troughs that trend northeast and curved troughs that trend in a northerly direction. This zone is most noticeable when figure 3-34 is viewed obliquely from the upper left-hand corner. The north-trending zone is on part of the surface that slopes gently toward the west, where the lower angle of lighting enhances the detectability of subtle breaks in slope.

The second set of linear features may be seen in the Hasselblad photographs as ridges, narrow troughs, and subtle tonal differences that trend in a northwesterly direction. These features can be seen best by viewing figure 3-34 obliquely from the upper right-hand and lower left-hand corners.

Approximately 1000 fragments between ¼ and 1 cm wide, 35 fragments between 1 and 6 cm wide, and two fragments 15 cm wide are scattered over the region studied. LRL specimen 10023 is the third largest fragment in the area. Most fragments of intermediate size are angular, but LRL specimen 10023 is rounded on the side facing the LM. One of the two largest fragments is very rounded, and the other has a prominent flat face. The small fragments are locally concentrated in low areas and are absent from higher areas. In addition, the number of small fragments tends to decrease toward the south and southeast.

Most larger fragments (1 to 20 cm wide) do not appear to be related to the depression in the surface. However, fragments measuring 1 to 3 cm wide that surround the two sharper depressions in the southwest part of the region could be pieces of projectiles that caused the two depressions. Impact of the larger rock in the northwest part of the region studied could have formed the depression that lies immediately to the right of the rock.

Most small fragments rest on the surface or are buried only slightly, as suggested by the relatively long, sharp shadows they cast. The intermediate and large fragments (1 to 15 cm wide) have fillets or embankments of fine-grained material around their visible sides.

On the two largest fragments and on LRL specimen 10023, the fillets appear to be better developed or perhaps restricted to the southeast faces. The fillets have gradational boundaries that change to sharp boundaries as they are traced from left to right around the southeast faces of the rocks. The distribution of the fillet is especially well shown on the large rock near the western edge of the region studied. Examination of stereophotographs AS11-40-5856 and AS11-40-5857 shows that the southeast face of the rock, which appears to overhang the lunar surface slightly, has a fillet that extends approximately two-thirds up the face and ends abruptly at the southwest corner of the rock.

These relationships suggest that the fillets were either formed from or enhanced by material moving toward the northwest, that is, parallel to the northwest set of lineations. Possibly both the fillets and the northwest-trending lineations have been partly formed from or enhanced by fine-grained material swept along the lunar surface by the effluent of the LM descent propulsion system.

Neither LRL specimen 10023 nor LRL specimen 10032 has any evident genetic relationship to the local depressions and craters or to either large or small fragments in the region studied. Like the other large fragments, LRL specimen 10023 has a well-developed fillet on the southeast face. If the fillets are of natural origin, the presence of the fillet on the LRL specimen 10023 indicates that the specimen was located in the position in which it was found for at least the period of time that was required for the fillet to be formed. However, if the fillets near the LM have been built largely during the landing by the descent engine effects, then LRL specimen 10023 could have been emplaced fairly recently.

Description. LRL specimen 10023 has an ovoid shape and is tapered at one end, with a broadly rounded side and a nearly flat side. The specimen

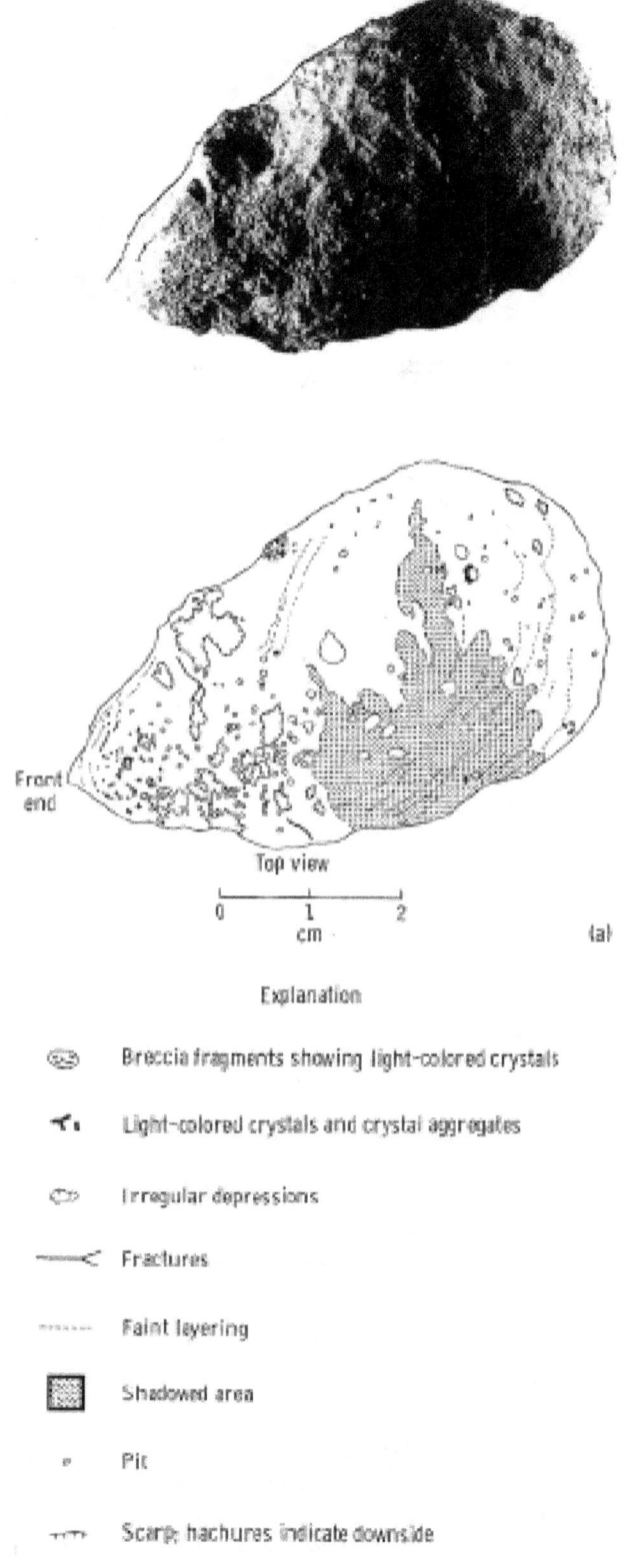

FIGURE 3-35. — Photographs and maps of LRL specimen 10023. (a) LRL photograph S-69-45394 and map showing the view from southwest of the specimen.

is approximately 5.5 cm long, 2 to 3 cm wide, and 1.5 to 2 cm thick. Part of the rounded side is covered with fine dust, but the flat side is fairly free of dust and reveals that the specimen is a fine-grained clastic rock with scattered subrounded rock fragments of up to 5 mm in diameter.

The rounded ovoid shape of the specimen (fig. 3–35(a)) is irregular in detail. In the central part of the rounded side, a broad depression is formed by many coalescing, shallow, and irregular cavities and round pits. Adjacent to this depression, toward the tapered "front" end, round deep pits are abundant and so closely spaced that some pits intersect others and indicate more than one generation of pitting. Fewer of these pits were evident near the edges of the specimen, and none could be identified from the photograph of the flat side of the specimen.

The flat side of the specimen (fig. 3–35(b)) is marked by two parallel flat surfaces separated by an irregular longitudinal scarp approximately ½ to 1 mm high. A few small cavities are present, but no round pits of the type found on the rounded side appear. An irregular fracture pattern occurs on the flat side. The fractures are short, discontinuous, and largely filled with dust.

On the rounded side of the rock, near the tapered end, is a set of short fractures, 3 to 9

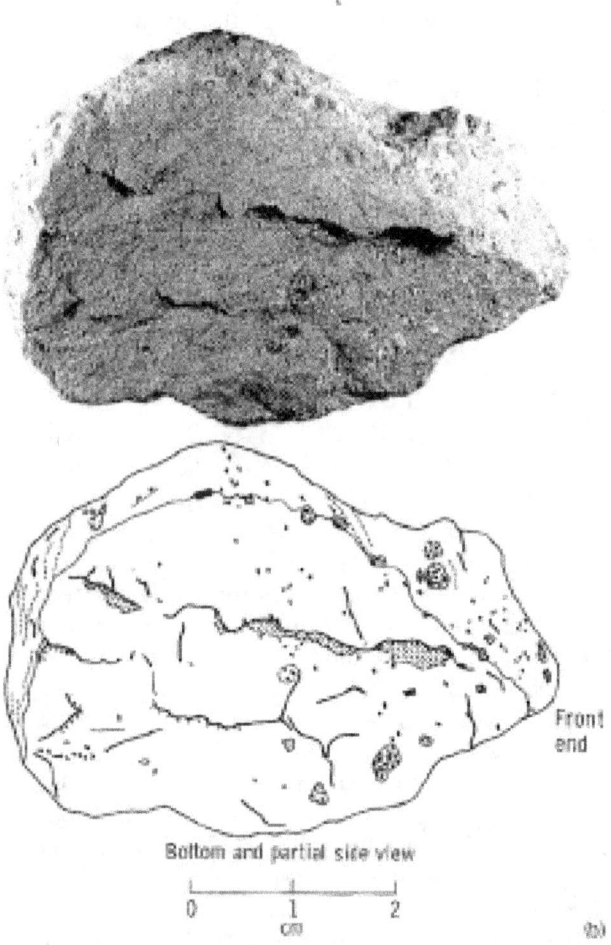

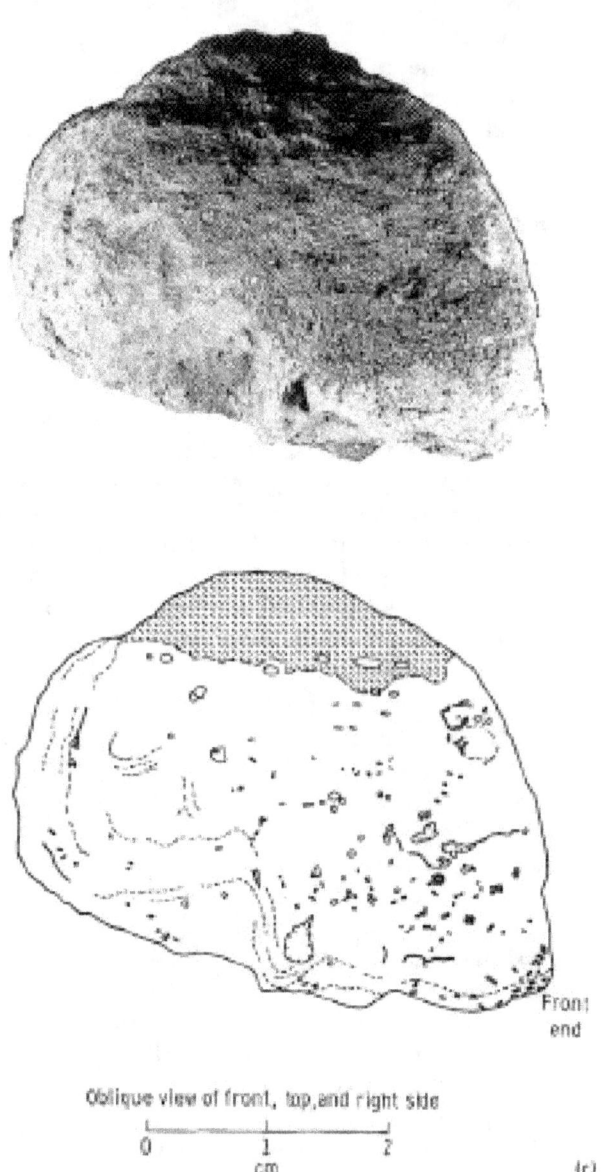

FIGURE 3–35 (continued). — (b) LRL photograph S-69-45393 and map of the flat side of the specimen.

FIGURE 3–35 (concluded). — (c) LRL photograph S-69-45387 and map showing the pits on the rounded side of the specimen.

mm long. These fractures are largely dust filled and do not appear to penetrate far into the rock. On a few sides and corners, short, curved fractures exist, which may be exfoliation features.

The numerous pits on the rounded side of the specimen (fig. 3-35(c)) range in size from approximately 0.3 to 2.3 mm. Most of these pits are deep, but some of the larger ones are shallow and bowl shaped. Many have raised rims, and some of the rims are surrounded by narrow, whitish halos. The larger pits grade in shape into irregular cavities and depressions. Cavities appear to be a composite of merged pits of various sizes. On the edges and flat side of the rock, the irregular cavities are few and small (mostly less than 1 mm long), and they appear to be caused by the dropping out of one or more lithic fragments or mineral grains.

LRL specimen 10023 is a fine-grained clastic rock containing widely and irregularly scattered, subangular rock fragments. The average grain size of the distinguishable fine material is approximately 0.2 mm. The rock fragments range from 0.5 to 4 mm and are mostly concentrated near the tapered end of the specimen, although fairly large fragments appear in other parts of the specimen. Most of the fine material consists of medium-gray, angular to subangular grains. Scattered through this matrix are white grains, many of them tabular or lath shaped, which resemble crystals and cleavage fragments. A few white grains appear to be curved and others are Y-shaped, which gives them the appearance of shards. These white grains range in size from 0.2 to 1 mm in length. Some of the larger laths show partial separation along apparent cleavage planes.

The larger rock fragments or clasts all seem to be fine-grained crystalline rocks with the texture of basalt. Randomly oriented, lath-shaped white "crystals" compose 40 to 60 percent of these rocks and are enclosed by a matrix of smaller gray minerals. The average grain size of the crystals is approximately 0.5 mm. Although many of the rock clasts are elongate, there appears to be no preferred orientation.

A faint layering that is essentially parallel to the flat face of the rock can be traced on the lower sides of the specimen. The layers are 1 mm or less in thickness and seem to be marked by alternating coarser and finer textures. Some of the elongate white grains appear to be concentrated along thin layers and are oriented parallel to the planar structure.

LRL specimen 10032 is bounded by numerous irregular faces that show no apparent control by joints or other platy structures. Most faces are remarkably fresh with only slightly rounded edges and corners. One face, however (fig. 3-36(a)), is smoothly undulating with more rounded edges and corners, and this face is therefore assumed to have been the top as the specimen rested on the lunar surface. This orientation is consistent with the correlation of LRL specimen 10032 with rock D in figures 3-19 and 3-22. The specimen has a pointed end, which is believed to have been oriented toward the LM landing site. No evidence remains of an apron or fillet of fine particles having been banked up against the sides of the specimen.

Irregular vesicles ranging in size from approximately 0.2 to 1 mm wide are unevenly scattered throughout the rock, but most are concentrated in two zones 5 to 6 mm apart, as shown in figure 3-36(b). Most of these vesicles show a parallel elongation and form a distinct lineation that is interpreted to represent the direction of flow of lava. Two larger, nearly spherical vesicles 1 and 2 mm wide appear on two of the faces. Most of the vesicles seem to be lined with smooth specular material, but in a few vesicles, small crystals are visible on the walls.

LRL specimen 10023 is a breccia of small subangular lithic fragments in an extremely fine-grained matrix. The specimen strongly resembles the material of the regolith, as photographed by the Apollo Lunar Stereo Closeup Camera, except that this specimen is strongly indurated. The shape and distribution of pits suggest that the rounded side of the specimen was formerly the top surface and that the flat side was the bottom surface. However, the specimen was not oriented with the flat side down at the time it was collected; both the rounded and the flat sides were partially buried in the regolith. Probably the rock had not been in its last position on the lunar surface for a substantially long period of time.

Description. LRL specimen 10032 is a small

GEOLOGIC SETTING OF LUNAR SAMPLES

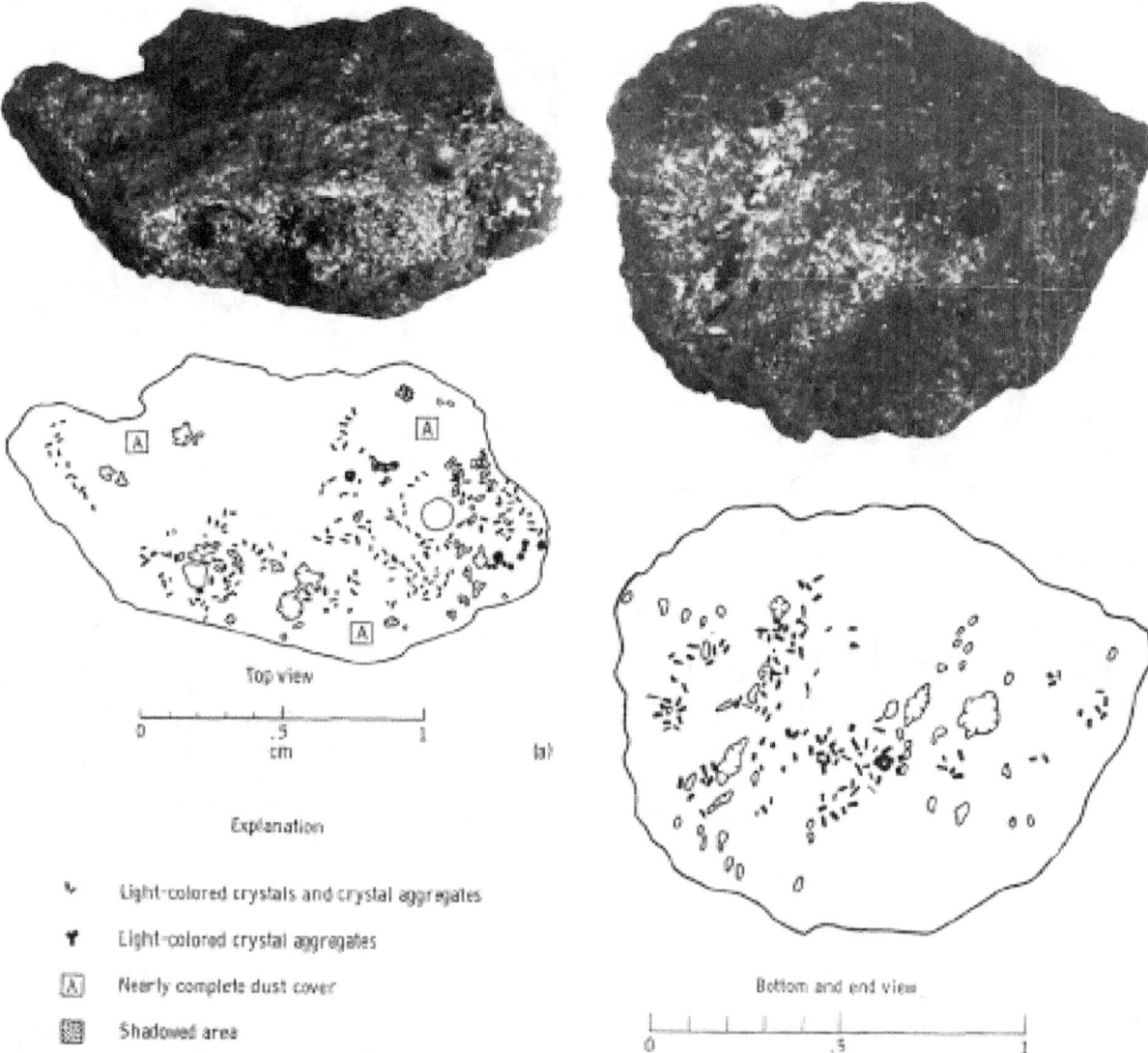

Explanation

- ᐯ Light-colored crystals and crystal aggregates
- ᵀ Light-colored crystal aggregates
- Ⓐ Nearly complete dust cover
- ▦ Shadowed area
- ⊘ Irregular pits
- • Small pits with white halos
- ⸴ Small vesicles (?)
- — Small fractures
- · Small, bright spherules

FIGURE 3-36. — Photographs and maps of LRL specimen 10032. (a) LRL photograph S-69-46010 and map of the top of the specimen.

FIGURE 3-36 (continued). — (b) LRL photograph S-69-46007 and map of the bottom and end of the specimen.

irregular rock fragment, approximately 2 cm long, 1 cm wide, and ¾ cm thick (the 12th largest from the contingency sample), whose location has been tentatively identified on the lunar surface. Study of the specimen has been based on LRL photographs S-69-46006 to S-69-46015, in which the specimen is enlarged approximately six times. The different views are stereoscopic.

On the assumed top of the specimen, a few scattered small, round pits appear that seem different from the vesicles. These pits are 0.2 to 0.4 mm wide and several are surrounded by a

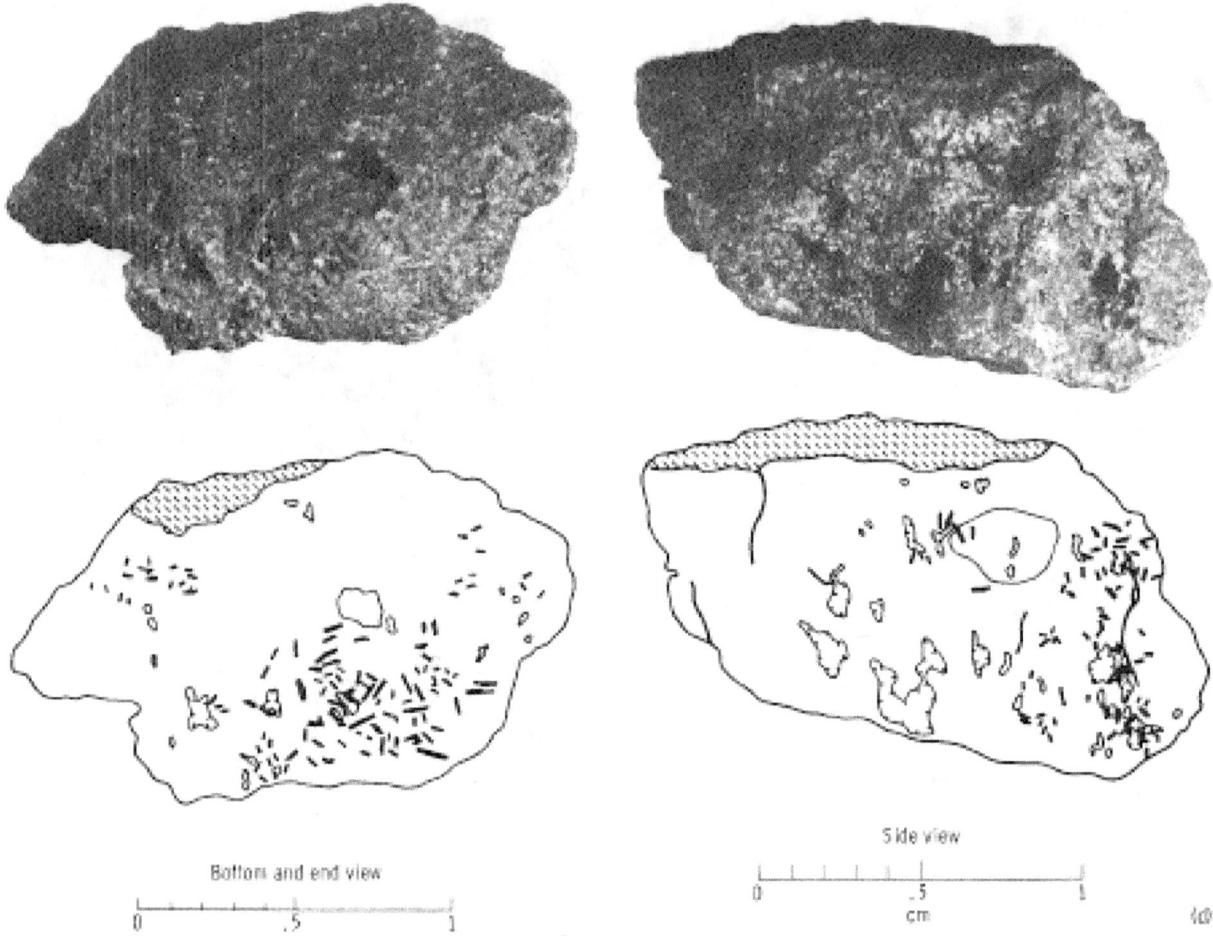

FIGURE 3-36 (continued). — (c) LRL photograph S-69-46014 and map of the bottom and end of the specimen.

FIGURE 3-36 (continued). — (d) LRL photograph S-69-46012 and map of the side of the specimen.

narrow, light-gray to white halo that is slightly raised in places. The origin of these pits is not clear, but their presence on only one face supports the interpretation that this face is the top side. The scarcity of pits of this type on the specimen, compared to pits of this type on the rounded tops of other returned lunar specimens, suggests that LRL specimen 10032 has been exposed for a short period of time.

LRL specimen 10032 is a fine-grained microcrystalline rock with a well-developed "diabasic" texture showing on some of the faces. The grain size averages between 0.1 and 0.2 mm. Abundant randomly oriented, lath-shaped, light-gray to white crystals resembling plagioclase form a feltlike network in a matrix of more equidimensional grains that are barely resolved in photographs that have been enlarged six times. The lath-shaped crystals, some of which are remarkably long and narrow, range from 0.2 to 1 mm long and from 0.1 to 0.2 mm wide. The longest crystals are shown best in figure 3–36(c). The matrix has a grain size of approximately 0.1 mm. Other views of LRL specimen 10032 are shown in figures 3–36(d) and 3–36(e).

The textural features of LRL specimen 10032 seem to clearly indicate an igneous origin, and the grain size, diabasic texture, and vesicle trains strongly suggest the derivation of the specimen from a lava flow, perhaps of diabase or diabasic basalt.

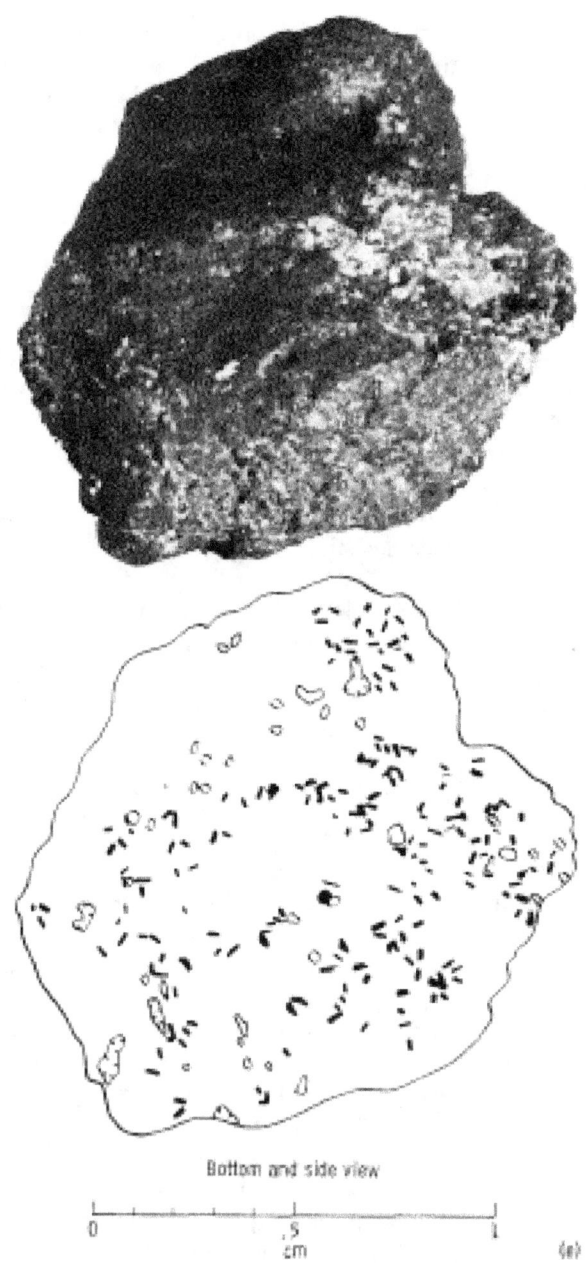

FIGURE 3-36 (concluded).—(e) LRL photograph S-69-46008 and map of the bottom and side of the specimen.

LRL Specimen 10046

Geologic environment. LRL specimen 10046 was collected in the bulk sample and has been identified in photographs of the lunar surface. Photographs AS11-40-5453, AS11-40-5738, AS11-40-5761, AS11-40-5762, AS11-40-5790, and AS11-40-5853 were taken from the LM windows. These photographs provide stereoscopic coverage but only with a very short stereoscopic base.

A region approximately 1.2 by 2.0 m surrounding LRL specimen 10046 was studied, and the region was mapped from photograph AS11-40-5853 (fig. 3-37). The rock was collected approximately 4 m west-southwest of the center of the spacecraft +Z footpad. The sample site is on the northeast rim of the double crater located southwest of the LM landing site. The rim of the double crater, where LRL specimen 10046 was collected, is broadly convex and pockmarked with many small craters. The subdued raised rim of a 2-m-diameter crater lies 15 cm south of the sample site, and many smaller subdued craters 15 to 20 cm in diameter are present within several meters of the sample site.

Elongate troughs a fraction of a centimeter deep, ⅔ to 2 cm wide, and 4 to 30 cm long are prominent features in this locality. Two sets of troughs, which are similar in size, shape, and distribution, are present. One set of troughs appears to trend roughly north-northwest, and the other set of troughs appears to trend roughly west-northwest. (A more precise orientation of the troughs, considering local surface slopes, will be obtained from photogrammetric reduction of the photographs.)

Approximately 1300 fragments between ⅔ and 3 cm wide, 50 fragments between 3 and 8 cm wide, and eight fragments between 8 and 15 cm wide are scattered over the region studied. The largest rocks in the region studied comprise six angular and six rounded rocks or rock groups and LRL specimen 10046, which is subrounded. While the fragments smaller than 8 cm wide are randomly scattered over the surface, those larger than 8 cm wide are concentrated in the western two-thirds of the region studied (fig. 3-37).

Three of the fragments shown by figure 3-37 to be angular appear to be partly buried, and five angular fragments appear to rest on the lunar surface. As many as one-third to one-half of the rounded fragments shown appear to be buried. The LRL specimen 10046 is resting on the lunar surface. The extent to which the other, smaller fragments are buried varies uniformly from almost total burial to no burial.

All but one of the larger fragments have fillets of fine-grained material banked against the sides that are observable in the photographs. The fillets are approximately 1 to 2 cm high on the large fragments. No correlation between the height of the fillets and the angularity of the rocks has been established.

The relatively large, angular fragments appear from the LM window partial panoramas to be more abundant in the vicinity of the 2-m-diameter crater (upper left, fig. 3-37). The possibility exists that the angular blocks were ejected from within the regolith (from the 2-m-diameter crater) and that the rounded fragments have been exposed at the lunar surface for a longer period of time. It is also possible that the angular fragments are more resistant to lunar surface erosive processes than are the rounded fragments.

The LRL specimen 10046 is unique among the rocks in the small block field. Photographs and maps of this specimen are shown in figure 3-38. The specimen is only slightly embedded in the regolith (no more than 1 or 2 cm) and is the only rounded fragment that protrudes markedly above the surface. The specimen is the only large fragment of equant shape. The fillet extends approximately 1 cm high on the specimen. The lack of burial and low fillet of the specimen suggest that it has been in its present position for a short period of time. However, the large number of pits on the lunar top surface of the specimen (fig. 3-38(e)) suggests that the upper surface of the specimen has been subjected to lunar surface processes for a long period of time.

Description. LRL specimen 10046 is a large rounded rock that was collected in the bulk sample. Six stereoscopic orthogonal views of the specimen were photographed. Arbitrary numbers and relative orientations have been assigned to the respective faces as follows:

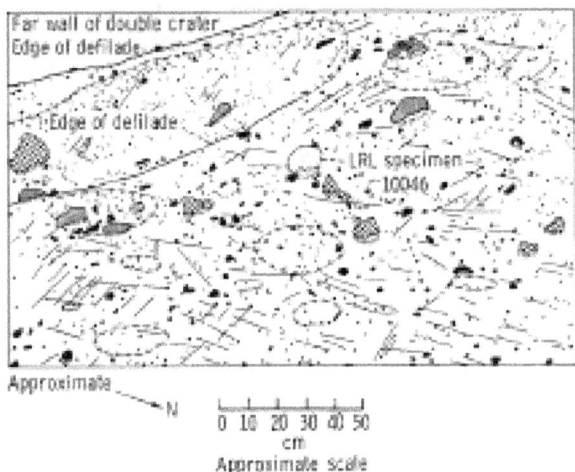

Explanation

- LRL specimen 10046
- Rounded rock fragment
- Angular rock fragment
- Rock fragment
- Fillet of fine-grained material on the side of a rock fragment; boundaries show the break in slope away from the fragment
- The break in slope at the margin of a subdued depression of a crater; hachures point downslope
- Axis of shallow trough

FIGURE 3-37. — Hasselblad photograph AS11-40-5853 and map showing the region around LRL specimen 10046 that was studied.

Face no.	Figure no.	Photographs
1 (front)	3-38(a)	S-69-45614 to S-69-45616
2 (right side)	3-38(b)	S-69-45611 to S-69-45613 and S-69-45623 to S-69-45625
3 (rear)	3-38(c)	S-69-45609 and S-69-45610
4 (left side)	3-38(d)	S-69-45617 to S-69-45619 and S-69-45629 to S-69-45631
5 (top)	3-38(e)	S-69-45620 to S-69-45622
6 (bottom)	3-38(f)	S-69-45626 to S-69-45628

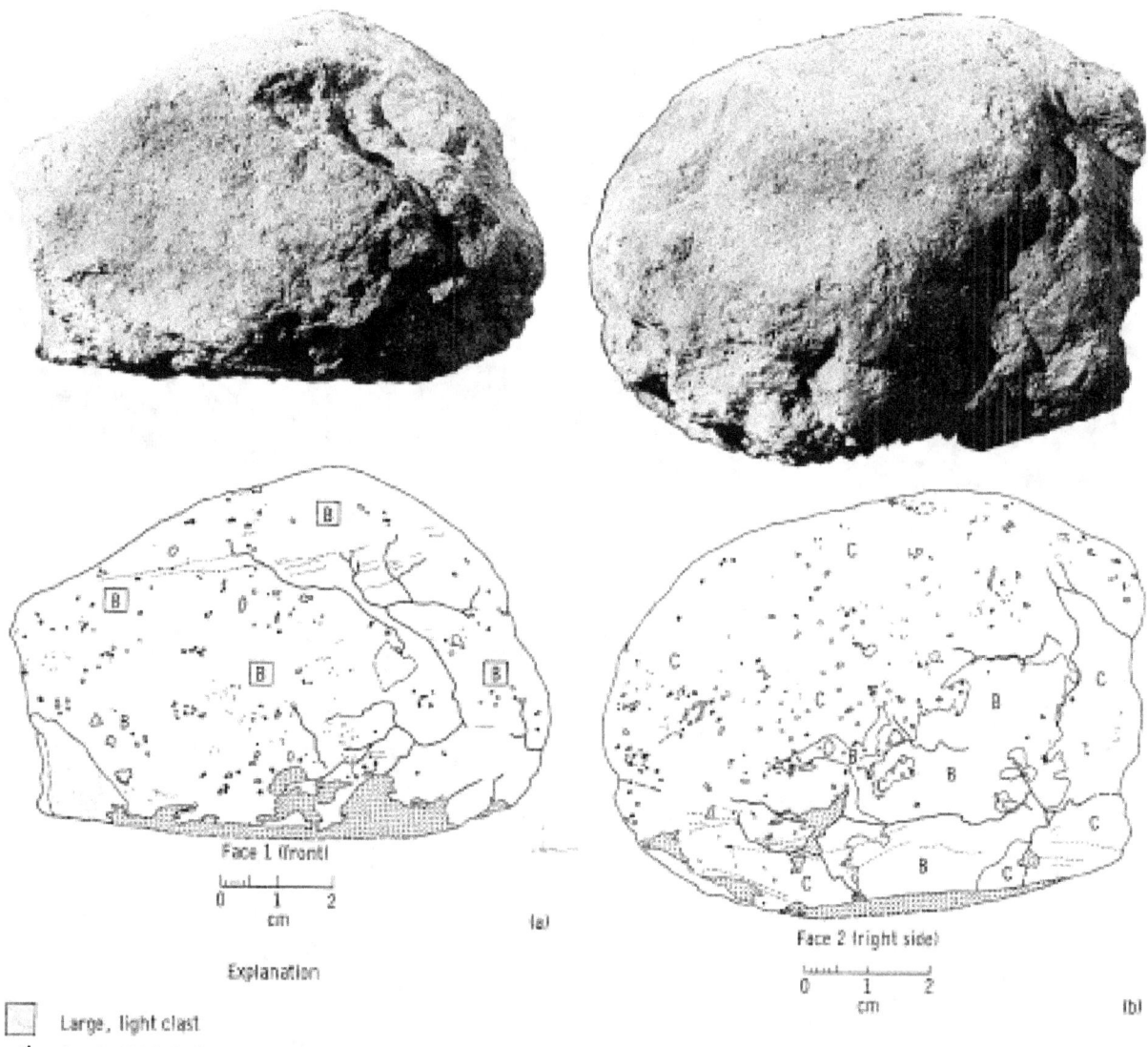

Explanation

	Large, light clast
	Small, light clasts
A	Heavy dust coating on the surface
B	Moderate dust coating on the surface
C	Light dust coating on the surface
	Shadowed area
----	Fracture
-----	Subtle trough or base of subtle scarp; probably the eroded edge of a resistant internal layer; in places, light-colored clasts are alined along adjacent ridges
⌒	Irregular cavity
○	Pit, commonly with raised rim and light-colored halo

FIGURE 3-38. — Photographs and maps of LRL specimen 10046. (a) LRL photograph S-69-45616 and map of the front (face 1) of the specimen.

FIGURE 3-38 (continued). — (b) LRL photograph S-69-45613 and map of the right side (face 2) of the specimen.

LRL specimen 10046 was oriented before collection (fig. 3-28) with face 1 to the southwest, face 2 to the southeast, face 3 to the northeast, face 4 to the northwest, and faces 5 and 6 as the top and bottom, respectively. Fine particles partly obscure faces 1, 3, 4, and 6.

LRL specimen 10046 is roughly ovoid, measures approximately 10 by 8 by 7.5 cm, and has three approximately orthogonal flattened sides (faces 3, 4, and 6). Face 1 is a relatively smooth, convex surface with a deep irregular indentation in the upper right corner. Face 2 is smoothly

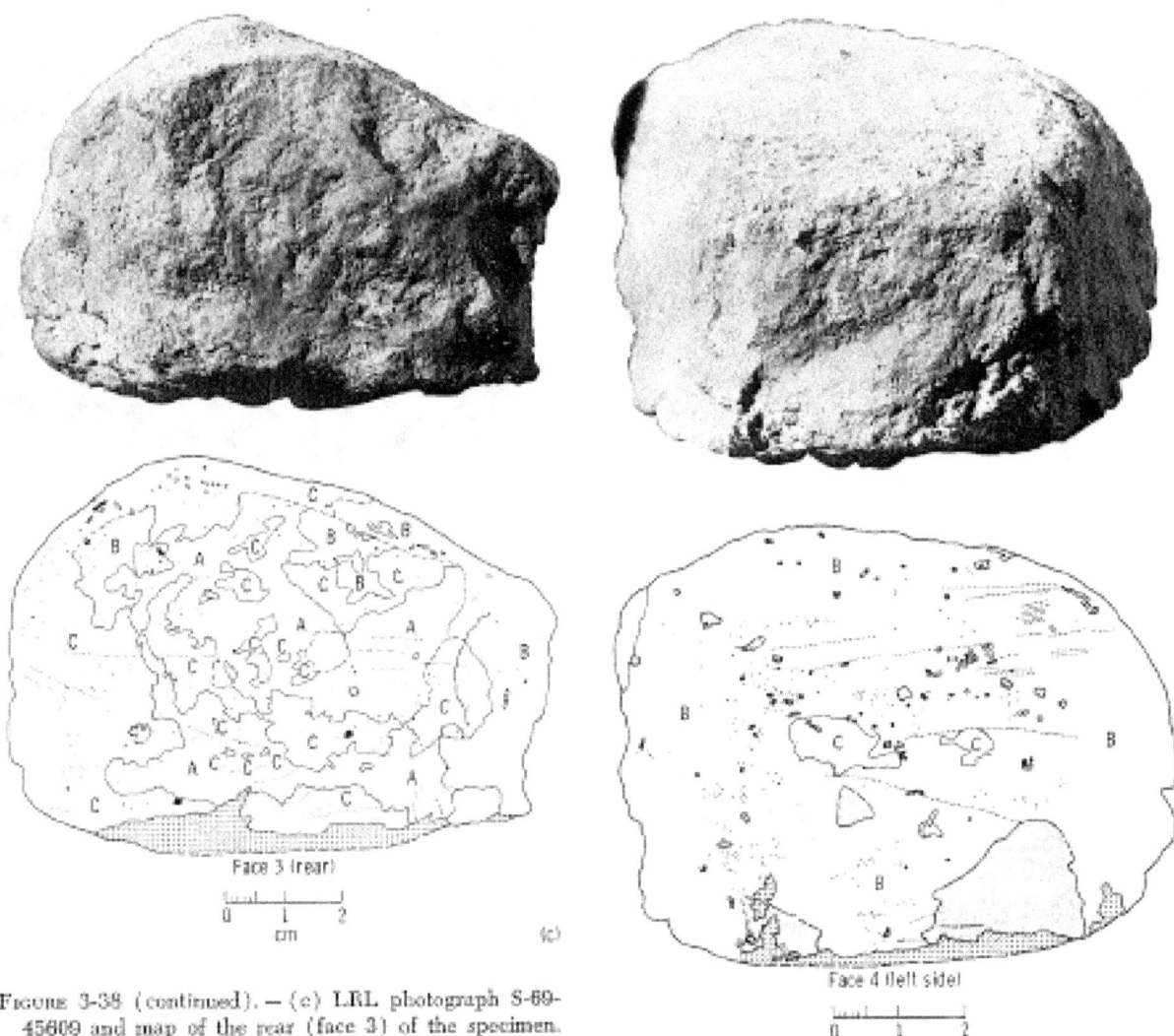

FIGURE 3-38 (continued).—(c) LRL photograph S-69-45609 and map of the rear (face 3) of the specimen.

rounded on the top and left sides and is irregularly broken at the bottom and the right sides. Face 3 is virtually flat but has sharp irregular indentations scattered over it. Face 4 is a rough, broken surface that is largely coated with fine particles. The lunar top surface of the specimen (face 5) is gently domed, smooth on a coarse scale, and densely pitted on a fine scale. The lunar bottom surface of the specimen is virtually planar and is heavily covered with fine particles.

Subtle troughs and ridges, which are approximately parallel with the base of the specimen, are visible on faces 1, 2, 3, and 4. These troughs and ridges are probably the eroded edges of a series of resistant layers that are most easily

FIGURE 3-38 (continued).—(d) LRL photograph S-69-45619 and map of the left side (face 4) of the specimen.

visible where the rock surfaces are obliquely lighted. The stratiform appearance of the specimen is enhanced by a very slight sorting of fine and coarse clasts in alternate layers and by the discontinuous local alinement of light-colored clasts.

A fracture system, which is most strongly developed on the lunar south and is displayed on faces 1, 2, and 6, cuts the specimen. One series of fractures can be traced approximately parallel with the base across face 2. Another set of arcu-

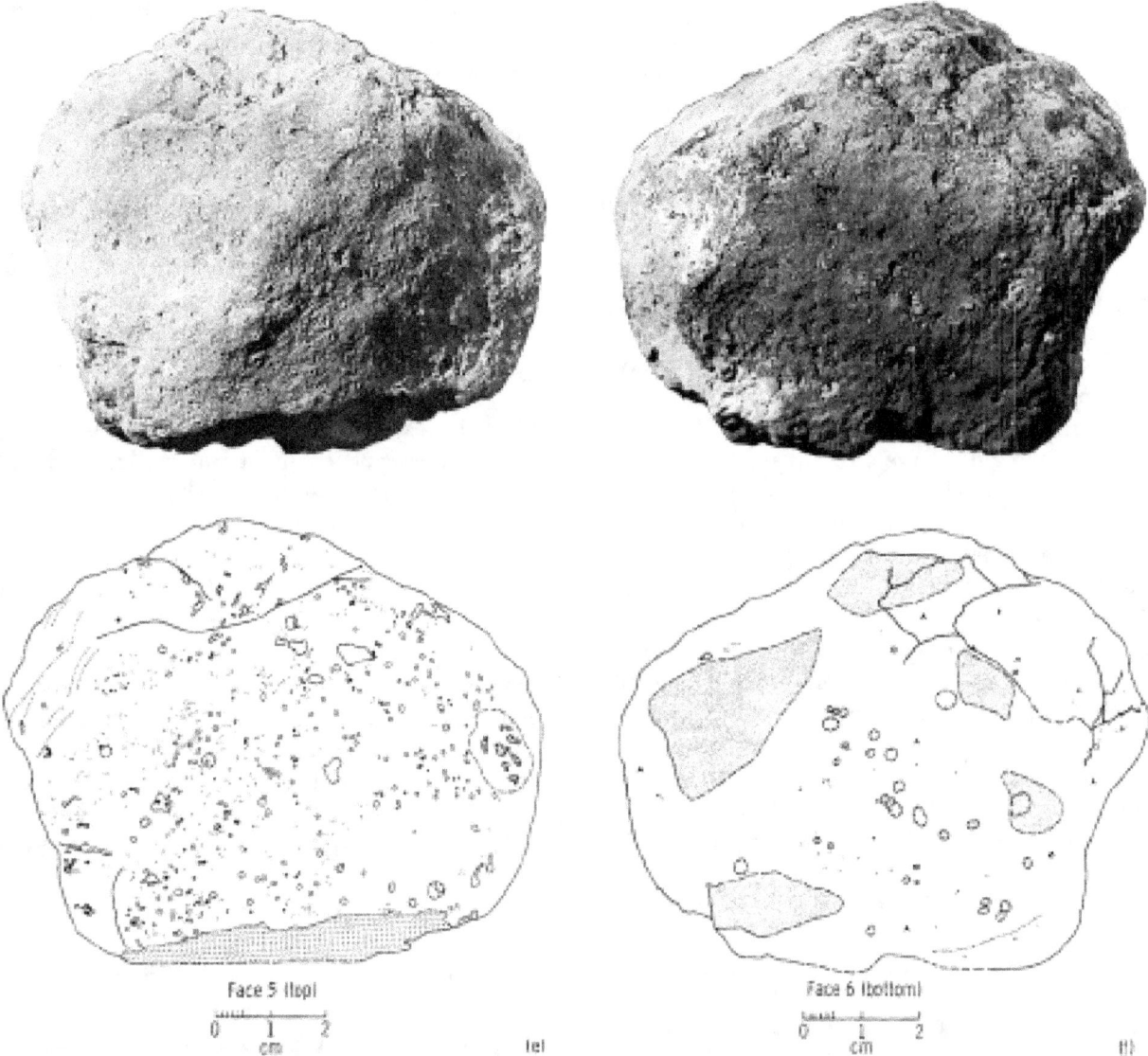

FIGURE 3-38 (continued).—(e) LRL photograph S-69-45622 and map of the lunar top (face 5) of the specimen.

FIGURE 3-38 (concluded).—(f) LRL photograph S-69-45628 and map of the lunar bottom (face 6) of the specimen.

ate fractures oriented from vertical to almost horizontal can be seen in faces 1 and 2. The orientation of the two arcuate fractures concentric with face 2 suggests that the fractures may be the inner boundaries of exfoliation shells.

Shallow circular pits that occur on the lunar top surface of the specimen appear less frequently on other faces of the specimen. The pits range in size from less than ½ to 3 mm in diameter. Most of the pits have a specular lining, which is generally of either a botryoidal or a radially ridged form. The pits commonly have raised rims with subdued to sharp profiles. Raised rims are commonly bounded by steep or reversed (undercut) outer slopes. Some of the sharp-rimmed pits have light-colored halos with diameters four times as great as the pit.

The pits are unevenly distributed on the six faces of LRL specimen 10046. They are most abundant on the lunar top surface of the specimen and occur with decreasing abundance on faces 2, 1, 6 (bottom), 3, and 4. On the top surface, the pits are closely spaced, and some pits are superimposed on other pits. In some cases, old pits are almost totally destroyed by superposition of younger pits. On the top surface and on faces 1 and 2, the number of pits increases rapidly with decreasing diameter between a 3-mm and a ¼-mm pit diameter. Too few pits are visible on faces 3 and 4 to judge the size-frequency distribution. On the bottom, which is largely coated with fine particles, only large pits are easily detected. Pits of a given size tend to cluster; that is, several pits of approximately the same size and degree of freshness tend to be found in proximity.

Irregular cavities are present on faces 2, 3, 4, and 5. These cavities range in size from 1 to 4 mm on the surface of the specimen and vary from very shallow to very deep. In some cases, the bottom and sides of the cavities are in shadow in all available photographs. The edges of these cavities range from irregular to smooth and from sharp-edged to subdued. The visible interiors of the cavities are commonly specular. On faces 2 and 4, some of the deep cavities are rounded in outline and interconnected. In several cases, the cavities are alined and tapered as though drawn out. Most of the irregular cavities occur in the rock matrix, but a few occur in large lithic clasts. These latter cavities may be vesicles.

The surface in which greatest pitting is evident is bounded in part by a short scarp (possibly 1 mm high) that defines a fairly continuous cap or rind on the rock. A cross section of the cap is revealed in the scarp. The upper surface of the cap is lighter gray than the surrounding sides of the rock and, except for pitting, exhibits no throughgoing structures. In a cross section, the cap exhibits a fine-textured hackly appearance, and microfractures that occur in the cap do not persist into the rock below. Major fractures penetrating the body of the rock do not cut the cap.

The rock is an aggregate of clasts consisting of approximately 10- to 20-percent recognizable irregular lithic fragments and possibly crystal fragments in a very fine-grained heterogeneous matrix. The average grain size of the matrix is less than 0.1 mm. The matrix consists predominantly of medium-gray grains with scattered light- and dark-gray clasts. Light-gray angular clasts occur bimodally both within the matrix and as coarse clasts. As a matrix constituent, these clasts range in size from less than 0.1 mm to approximately 1.5 mm. Coarse clasts of light-gray material range in diameter from approximately 5 mm to 3 cm.

The coarse, light-gray clasts are angular and irregular but are equant or rarely tabular in form. The clasts have sharp contacts with the matrix. Some of the finer clasts are dull, some are highly reflective, and some appear to be fragments of single crystals. One fragment on face 1 resembles a zoned feldspar with an altered core and twinned rim. At least two types of rock are represented in the larger light-gray clasts. One is a very fine-grained vesicular rock that is well represented by large clasts on face 4. The other rock is crystalline with an average grain size of approximately 0.3 mm.

Dark-gray clasts are scattered throughout the matrix but are the least abundant of the three photometrically distinguishable components. Individual grains range from less than 0.1 to 0.2 mm, but locally they are spatially concentrated to form an irregular, subrounded aggregate of several millimeters in width (lower center of face 1).

In a few cases, light-gray clasts stand in relief above the surrounding matrix, which suggests that these clasts are more resistant to processes of erosion occurring on the Moon. In contrast, aggregates of dark-gray clasts (face 1) appear to erode in negative relief.

LRL specimen 10046 is a microbreccia. The general character and distribution of its component fragments are similar to those exhibited by photographs of the unconsolidated surface material taken by the lunar closeup camera. This similarity suggests that the specimen may have been derived by a process of induration of the unconsolidated surface material. The subtle stratification that is visible on three faces of the rock may have been produced by the process of induration, or it may indicate a sedimentary layering process. The external form of the specimen has probably been influenced by at least four separate processes: (1) fracturing of the body rock, (2) exfoliation, (3) pitting, and (4) surface erosion (or ablation).

LRL specimen 10046 has a complex history of movement on the lunar surface, and this history is reflected in its surface form. The most angular face (face 3) is oriented toward the northeast, and the least angular face (face 6) is on the bottom. The lack of pitting on face 3 indicates that this face was probably the bottom of the specimen at one time, leaving faces 1 and 6 exposed to processes of pitting. At some time in the recent past, possibly at the time one of the nearby craters was formed, the rock was reoriented. This reorientation exposed face 3 to erosion. The general rounding of the surface, the fracturing of the body of the rock, and the fact that three faces of the rock are relatively angular suggest that the reorienting process may have happened more than once and that the process may have happened at irregular intervals of time.

LRL Specimen 10047

Description. LRL specimen 10047, which is a small rock from the bulk sample box, was collected within 8 m of the LM from a region on either the northwest or the southwest side of the LM landing site. The specimen has not yet been identified in photographs of the lunar surface. The specimen has an unusually angular shape, except for one side that appears to have been rounded off by lunar erosional processes, and the rounded side is assumed in this description to be the top side. The maximum dimensions are approximately 5.5 cm in length, 4 cm in width, and 3.3 cm in thickness. Although much of the specimen is covered by a thin, dark-colored coating of fine particles, a finely crystalline texture shows on parts of most faces, and a faint but definite primary foliation shows on the top and one end of the specimen.

This study is based on LRL photographs S-69-45632 to S-69-45655. A total of 24 photographs, in which the rock was enlarged approximately three times its original size, gave excellent stereoscopic coverage and good definition of the structural and textural features (fig. 3-39).

As viewed from the top (LRL photographs S-69-45644 to S-69-45646), LRL specimen 10047 is roughly rectangular in shape. This shape is apparently controlled by a conjugate system of small, irregular joints; but the unusual angu-

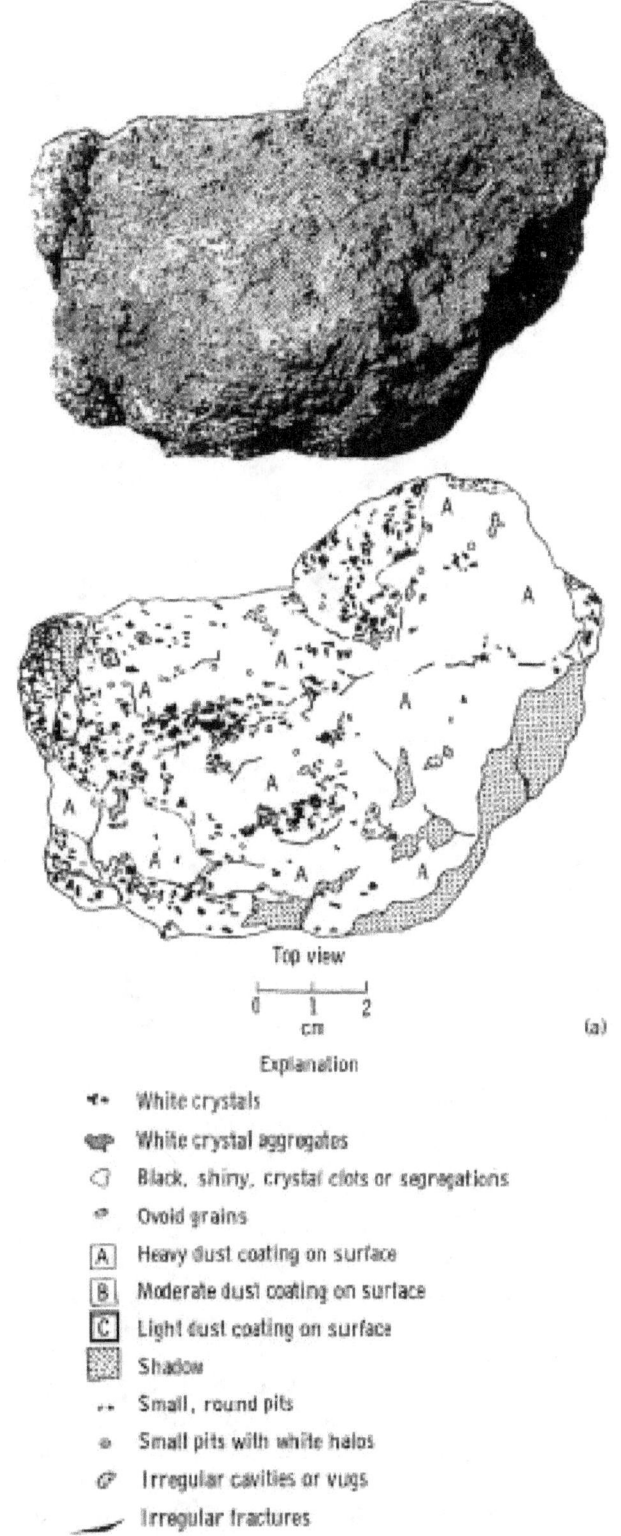

FIGURE 3-39. — Photographs and maps of LRL specimen 10047. (a) LRL photograph S-69-45644 and map of the top of the specimen.

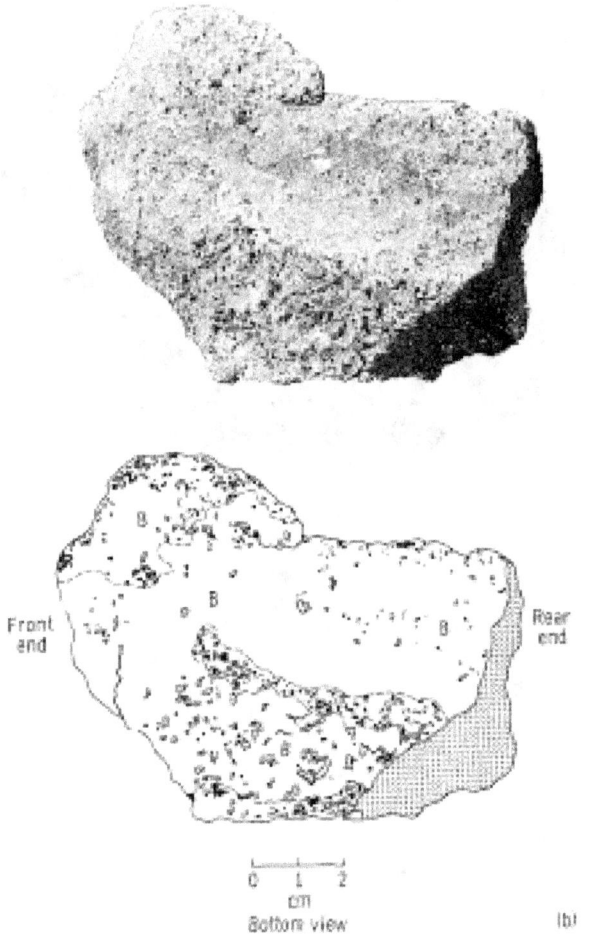

FIGURE 3-39 (continued). — (b) LRL photograph S-69-45649 and map of the bottom of the specimen.

larity (exhibited on the other faces) is caused by another system of joints that intersects the first system at angles of 20° to 45° and by the occurrence of irregular fractures. The bottom of the specimen (LRL photographs S-69-45649 to S-69-45651), with faces meeting at an angle of approximately 70°, exhibits the most angular edge. This edge may have been embedded in the regolith to a depth of 1 to 2 cm. Other edges are considerably more rounded, particularly those that bound the top surface.

Two types of small fractures appear on various faces, particularly on the rounded top of the rock. One type consists of fairly straight, short (1 to 2 cm) fractures that are roughly parallel to the faces of the rock. Most fractures

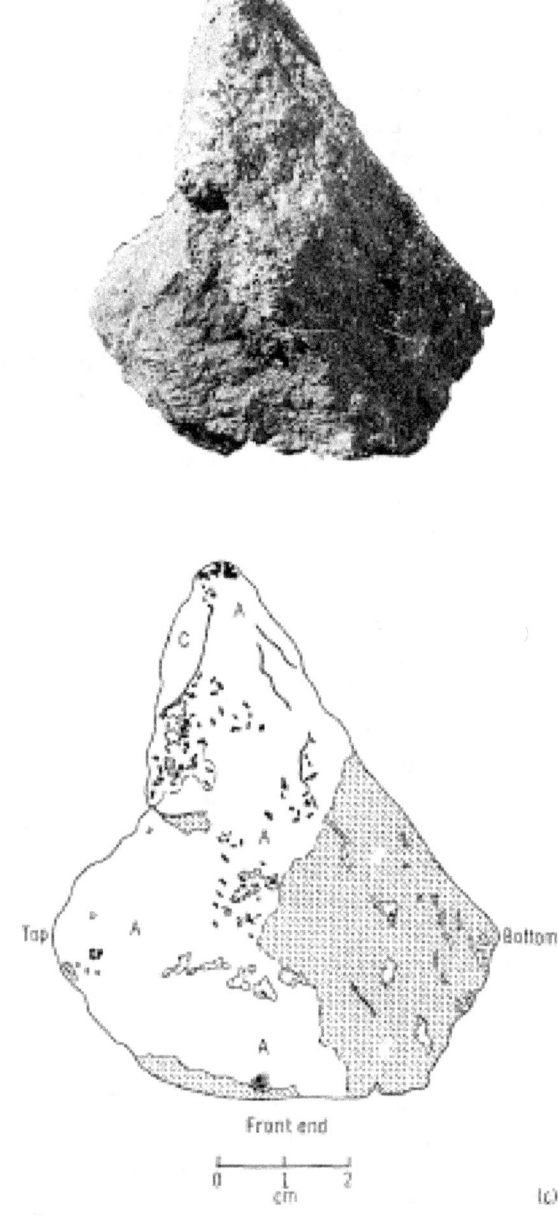

FIGURE 3-39 (continued). — (c) LRL photograph S-69-45647 and map of the front of the specimen.

are partially filled with fine particles. The second type consists of curved fractures that are roughly parallel to the rounded ends and edges of the rock. In detail, these fractures are irregular, partly open, and bordered in part by crumbly granular material. The fractures appear to bound

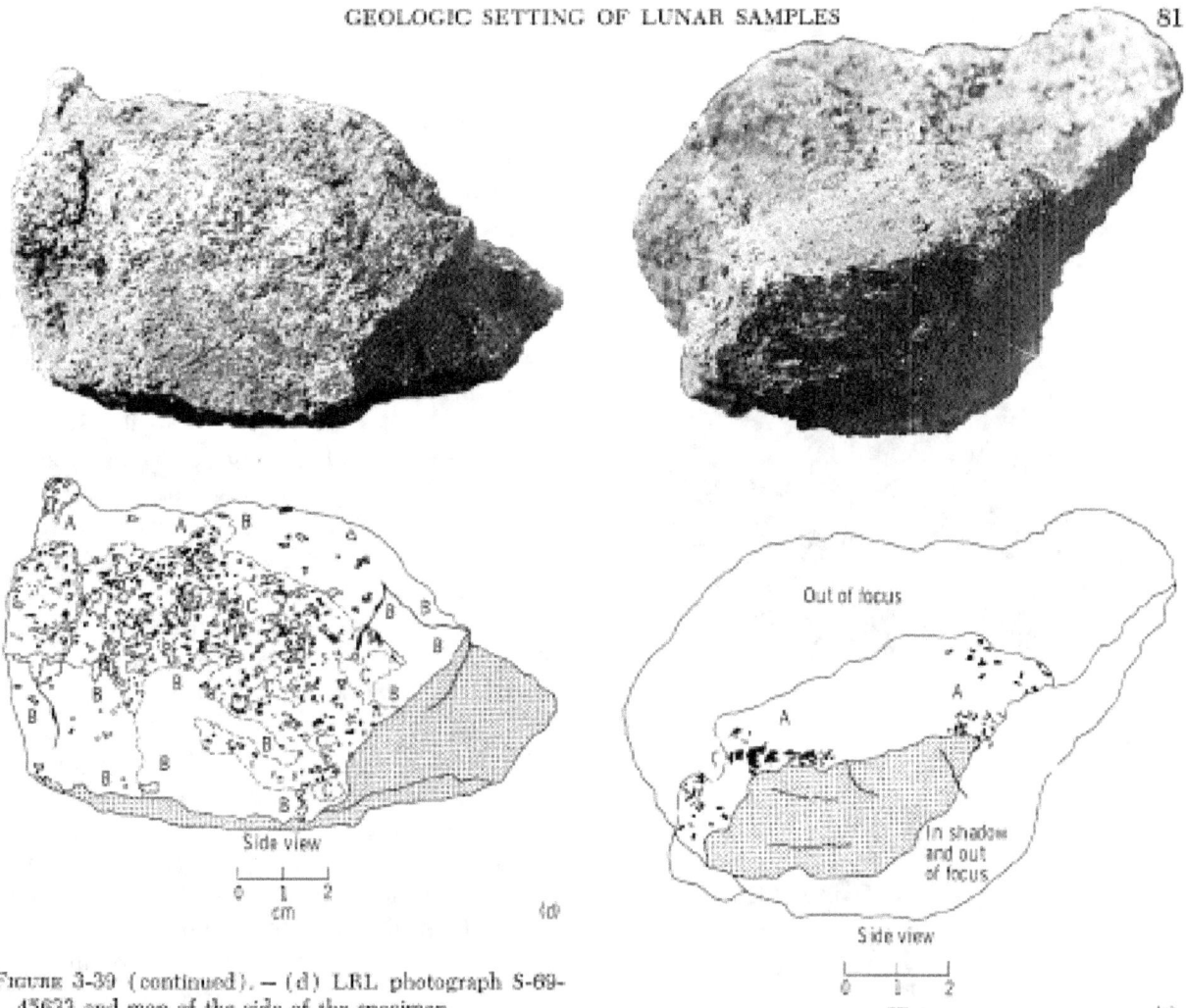

FIGURE 3-39 (continued). — (d) LRL photograph S-69-45633 and map of the side of the specimen.

FIGURE 3-39 (continued). — (e) LRL photograph S-69-45638 and map of the side of the specimen.

a series of exfoliation shells that are most prominent on the rounded top.

Scattered over the various faces of the specimen are irregular cavities or vugs that are 0.5 to 2.5 cm long. The cavities are most abundant on one of the bottom faces. Small crystals can be seen projecting from the cavity walls at various angles. Because of their extreme irregularity in shape and distribution, these vugs do not appear to be ordinary vesicles; they resemble cavities formed by leaching or by removal of some unknown mineral or mineral aggregates. The crystals projecting from the walls seem to be integral to the fabric of the rock, rather than having been deposited in the vugs.

Small pits 0.2 to 1.5 mm in diameter are irregularly scattered over the top surface of the specimen, and a few pits appear on the sides of the specimen. Some of the pits are steep sided as though drilled, while others are conical with raised rims. Several of the pits are surrounded by a white halo that seems to be composed of an array of white mineral grains with a slightly raised outer margin.

The texture of LRL specimen 10047 is best seen on a side face that is shown by LRL photographs S-69-45632 to S-69-45634 and is sketched on LRL photograph S-69-45633. The rock appears to be holocrystalline with an average grain size of approximately 0.5 mm, although some crystals are slightly more than 1 mm in length. At least two components can be recognized. White, lath-shaped to stubby rectangular crys-

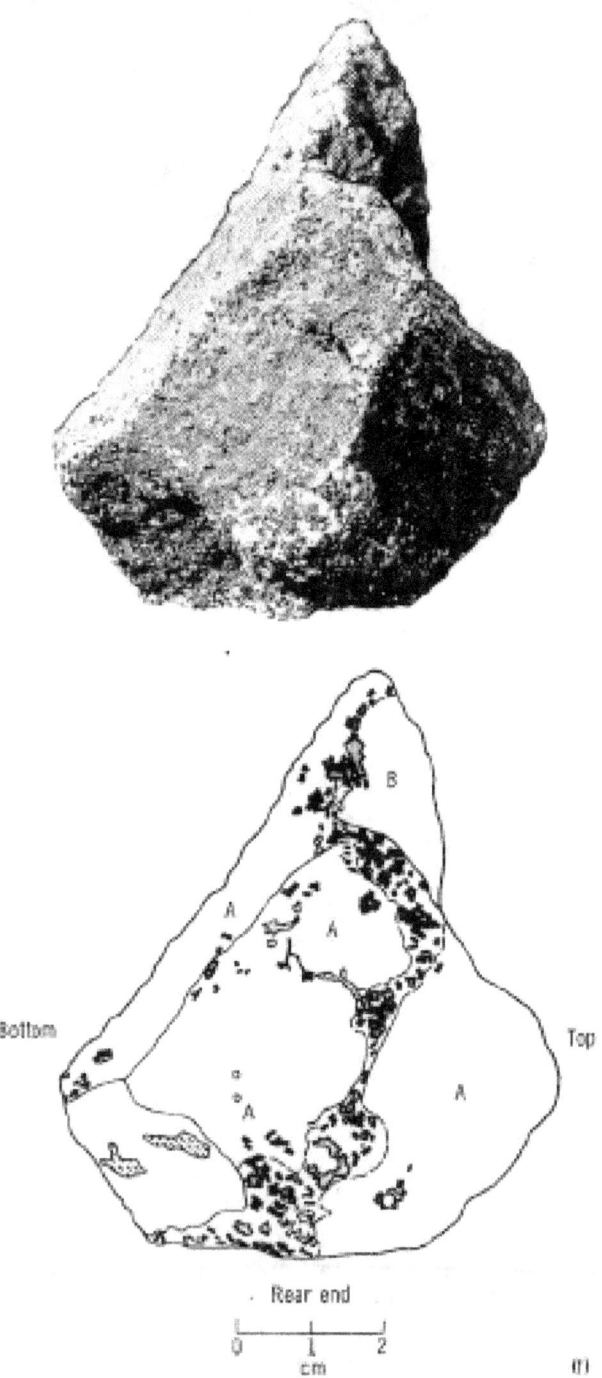

FIGURE 3-39 (concluded). — (f) LRL photograph S-69-45652 and map of the rear of the specimen.

large as 2 mm in diameter. Some of these aggregates exhibit a faint cleavage that is typical of plagioclase, especially on a relatively fresh broken surface such as that shown in LRL photograph S-69-45561.

Surrounding the white crystals is a more finely granular matrix of dark-gray to nearly black minerals of approximately 0.2- to 0.5-mm grain size. Many of these minerals appear as stubby subhedral prisms that exhibit brightly reflectant crystal and cleavage faces. Locally, these dark grains are grouped in aggregates or clots up to 2 mm in diameter. Whether the dark grains represent one mineral or more than one mineral cannot be discerned from the photographs.

Fresh broken surfaces shown in LRL photograph S-69-45561 exhibit, on the top and upper sides, a thin surface crust approximately 1 mm or less thick of high albedo, in contrast to the darker tone of the fresh interior. The extreme whiteness that characterizes the surface of the crystals, resembling plagioclase, and the relatively high reflectivity of other grains are apparently caused by a lunar weathering process. Shock metamorphism, a process by which the minerals on the surface may have been minutely shattered by the impacts of many small particles on the rock, seem to be the most reasonable explanation for the formation of this crust. Shock effects may also account for the white halos that are evident around many of the round pits.

On the rounded top and on one end of LRL specimen 10047, the white crystals and aggregates exhibit a crude but distinct foliation, which trends roughly parallel to the long dimension of the top surface and dips at a steep angle to this surface. Many of the lath-shaped crystals are oriented almost parallel with the foliation and, along with crystal aggregates, are concentrated in crude layers. This foliation is similar to primary flow structures in both extrusive and intrusive terrestrial igneous rocks.

The textural features and general appearance of LRL specimen 10047 strongly suggest an igneous origin, but the medium-grain size could be characteristic of either a thick lava flow or a small intrusive body such as a dike. The faint foliation indicates the small intrusive body to be the better choice. The specimen resembles a terrestrial diabasic basalt or fine-grained gabbro.

tals resembling plagioclase make up approximately 50 percent of the rock. The distribution of the crystals gives the rock a slightly diabasic texture. The crystals range from 0.3 to 1.2 mm long, and many are grouped into aggregates as

References

3-1. MILTON, D. J.: Geologic Map of Theophilus Quadrangle of the Moon: Geologic Atlas of the Moon, Scale 1:1,000,000. U.S.G.S. Map I-546 (LAC 78), 1968.

3-2. HEACOCK, R. L.; KUIPER, G. P.; SHOEMAKER, E. M.; UREY, H. C.; and WHITAKER, E. A.: Interpretation of Ranger VII Records: Ranger VII, Part II, Experimenters' Analyses and Interpretations. TR 32-700, Jet Propulsion Laboratory, Calif. Inst. Tech., 1965, pp. 9-73.

3-3. MCCORD, T. B.: Color Differences on the Lunar Surface. J. Geophys. Res., vol. 74, no. 12, 1969, pp. 3131-3142.

3-4. MULLER, P. M.; and SJOGREN, W. L.: Mascons: Lunar Mass Concentrations. Science, vol. 161, 1969, pp. 680-684.

3-5. SHOEMAKER, E. M.; MORRIS, E. C.; BATSON, R. M.; HOLT, H. E.; LARSON, K. B.; MONTGOMERY, D. R.; RENNILSON, J. J.; and WHITAKER, E. A.: Television Observations from Surveyor. TR 32-1265, Jet Propulsion Laboratory, Calif. Inst. Tech., 1969, pp. 21-136.

4. Apollo 11 Soil Mechanics Investigation

N. C. Costes, W. D. Carrier, J. K. Mitchell, and R. F. Scott

The Apollo 11 lunar landing mission afforded man the first opportunity for direct collection of data relating to the physical characteristics and mechanical behavior of the surface materials of an extraterrestrial body by other than remote means. In particular, the first manned lunar landing provided a unique capability for acquiring information that would aid in the accomplishment of the following broad objectives:

(1) To enhance the scientific understanding of the nature and origin of the materials, and the mechanisms and processes responsible for the present morphology and consistency of the lunar surface

(2) To provide engineering data on the interaction of manned systems and manned operations with the lunar surface, thereby aiding in the evaluation of the Apollo 11 mission, and in the planning of future lunar surface scientific investigations and related engineering tasks supporting these activities

To obtain this information, the Apollo Program Office directed that a Soil Mechanics Investigation be included in the scientific experiments planned for the Apollo 11 mission. The investigator team was charged with the responsibility for the systematic acquisition and analysis of lunar soil mechanics data to insure maximum return of geotechnical information from each facet of the planned lunar surface operations.

Specific scientific objectives of the Soil Mechanics Investigation at the Apollo 11 landing site included the following:

(1) To verify lunar soil models previously formulated from Earth-based observations and laboratory investigations and from lunar orbiting and unmanned lunar landing missions

(2) To determine the extent of variability in lunar soil properties with depth and lateral position

(3) To aid in the interpretation of geological observations, sampling, and general documentation of maria features

In addition, the Soil Mechanics Investigation was aimed toward the following engineering objectives:

(1) To obtain information relating to the interaction of the lunar module (LM) with the lunar surface during landing and to lunar soil erosion caused by the spacecraft engine exhaust

(2) To provide a basis for altering mission plans because of unexpected conditions

(3) To assess the effect of lunar soil properties on astronaut and surface vehicle mobility

(4) To obtain at least qualitative information needed for the deployment, installation, operation, and maintenance of scientific and engineering stations and equipment to be used in extended lunar exploration

Knowledge of Lunar Surface Mechanical Properties Before Flight

No attempt will be made in this section to give a complete history of the development of the state of knowledge of the lunar surface mechanical properties. The purpose of this discussion is to provide a background for better evaluation of the results of the mission and to make comparisons between the properties that have been deduced from various sources. Using these comparisons, the reliability of different sets of measurements can be assessed, and the suitability of the techniques used prior to Apollo 11 can be evaluated.

Several sources of preflight information about lunar surface mechanical properties exist. These sources of information include ground-based visual observations, thermal measurements, radio and radar measurements, the lunar surface photographs obtained by the U.S. Ranger and Lunar Orbiter spacecraft, the data from landings of the Soviet Luna 9 and Luna 13 spacecraft, and the direct estimates of the lunar surface properties derived from the five soft landings of the U.S. Surveyor spacecraft.

The results of the Surveyor spacecraft tests and analyses led to the construction of a lunar soil model of an essentially incompressible, slightly cohesive soil that is composed primarily of grains ranging in size from silt to fine sand. The lunar soil behaved in a manner similar to the behavior of terrestrial soils with a density of approximately 1.5 g/cc. A cohesion of approximately 0.1 psi and a friction angle of 35° to 37° (in the normal pressure range of a few psi) satisfactorily represented the mechanical properties observations made of the lunar material to a depth of several inches. Where the soil extends to depths greater than several inches, an increase in strength with depth was observed. In places, the soil may overlie rock fragments that are only a few inches or less beneath the surface. (More information can be found in refs. 4–1 to 4–11.)

Postulated Soil Behavior During Descent and Touchdown of the LM

On the basis of the previously discussed lunar soil model, various soils have been postulated and used for calculations related to the descent and landing of the LM on the lunar surface. Some of these computations have been concerned with the dynamics of the landing;[1] for example, from computations, LM footpad penetrations of 4 to 6 in. were expected in a simultaneous four-point touchdown at a vertical downward velocity of 3 fps and at zero lateral velocity, if the lunar soil extended to a depth of one or two times the footpad diameter (i.e., 3 to 6 ft).

The effect of the descent propulsion system (DPS) engine exhaust on the soil surface has also been studied from several viewpoints. If the lunar surface is considered to consist of soil grains in the postulated size range, but is taken to be impervious to gas flow, calculations by Hutton (ref. 4–12) indicated the extent of lunar soil surface erosion that may be anticipated in a nominal descent. For the soil model considered, Hutton found that erosion would begin during a vertical descent to the surface when the exit plane of the DPS engine nozzle was approximately 25 ft above the surface. Hutton also found that following a vertical descent, the final erosional crater that would develop beneath the engine after landing and engine shutdown would be approximately 3 to 4 in. deep and 5 to 6 ft in diameter. This erosional crater formed by the flow of exhaust gases over the lunar surface would be ring shaped, and the maximum crater depth would occur some distance from the center of the crater. The effect of the blowing surface material on the visibility from the LM was also examined.

By considering the lunar soil to be a medium that is permeable to gas flow and has a permeability in the range appropriate for the grain size of the lunar soil model material and by ignoring the erosion mechanism, Scott and Ko (ref. 4–13) examined the mechanics of compressible gas flow through the soil medium under lunar surface conditions. On analysis of the Surveyor test results and of postflight tests, scaled to the LM, Scott and Ko found that a vertical descent (or steady engine firing in one position) followed by a rapid shutdown of the engine could give rise to gas pressures inside the soil that would exceed the lunar weight of the soil overburden. Thus, shutdown could be followed by a venting of the gas through the surface soil, accompanied by upward ejection of the surface soil. The extent and amount of soil removed by such explosive outgassing depend, for any given soil and engine, considerably upon the flightpath and the engine-shutdown pressure transients. A slow vertical descent and a rapid decay at shutdown would produce the largest quantity of ejected soil material.

Postulated Soil Behavior During Lunar Surface Extravehicular Activity

Calculations based upon the adopted lunar soil model indicated that during the lunar surface extravehicular activities (EVA), the astronauts' boots should not sink more than approximately 1 to 2 in. into the lunar surface if the lunar soil extended to a depth of a few inches to several feet. Traction on the lunar surface was

---

[1] E. M. Shipley, personal communication, 1968.

anticipated to be good. No difficulties in obtaining surface soil samples, driving core tubes, or installing staffs in the ground were expected if the soil was sufficiently deep. Mobility problems might be expected only if an astronaut attempted to descend or ascend crater walls with slope angles greater than approximately 15°.

Data Sources for the Investigation

The Soil Mechanics Investigation was included at a late phase of the Apollo 11 mission planning; consequently, the main guideline given to the investigator team was that no special soil mechanics testing or sampling devices were to be added to the hardware already planned for the mission. Accordingly, the following were the main sources from which soil mechanics data could be extracted:

(1) Real-time astronaut observations, descriptions, and comments during the mission

(2) Real-time television during the lunar surface EVA

(3) Sequence camera, still camera, and close-up stereoscopic camera photography

(4) Spacecraft flight mechanics telemetry data

(5) Various objects of known geometry and dead weight that came in contact with the lunar surface during the mission and subjected the lunar soil to either a static or dynamic loading; such objects included the LM, the astronauts, and the Early Apollo Scientific Experiments Package (EASEP) instrument units

(6) The Apollo lunar handtools

(7) Various poles and shafts inserted into the lunar surface in the course of the EVA, including the contingency sampler handle, the Solar Wind Composition (SWC) experiment staff, the flagpole, and the core tubes

(8) Astronaut technical, photographic, and scientific debriefings

(9) Preliminary examination of Earth-returned lunar soil and rock samples at the Lunar Receiving Laboratory (LRL)

Figure 4–1 shows the LM footpad geometry and the strut force-stroking characteristics. If the LM rests on a level lunar surface and the LM weight is equally distributed among the four footpads, each footpad exerts a static vertical force of approximately 700 lb on the lunar surface.

Figure 4–2 shows the configuration of the astronaut boot sole. In lunar gravity, the weight of the suited astronaut is approximately 64 lb. The area of a flat boot sole, not considering the ribs of the sole, is approximately 65 in.^{2}

Because the astronauts were essential in conducting the investigation and providing real-time evaluation, especially with regard to unexpected results, special effort was made to delineate, during astronaut training, simple tasks and observations yielding meaningful soil mechanics information, as well as expected behavior of lunar soil under various loading conditions.

The EASEP instrument units are shown in figures 4–3 and 4–4. In lunar gravity, the weights of the Laser Ranging Retroreflector (LRRR) and the Passive Seismic Experiment Package (PSEP) are 8.7 and 18.7 lb, respectively. The bearing areas of the pallets on which these two instrument packages rest are 708.5 and 732.1 in.^{2}, respectively. The Apollo lunar handtools are shown in figure 4–5. These tools are described in detail in the report concerning lunar geology. The bottom sections of the shafts mentioned in item 7 of the list of soil mechanics data sources are shown in figures 4–6 to 4–9.

In addition to the data obtained from the Apollo 11 mission, preflight data were available from simulation studies performed by the investigator team on soils having physical and mechanical characteristics similar to those indicated by the results obtained from the Surveyor missions. These test results have been documented and reports are being prepared by N. C. Costes et al. (George C. Marshall Space Flight Center), J. K. Mitchell et al. (University of California, Berkeley), and R. F. Scott and T. D. Lu (California Institute of Technology, Pasadena). In addition to laboratory tests performed to determine the basic mechanical properties and physical characteristics of the simulated lunar soil, an effort was made to ascertain the behavior of the material under conditions expected during EVA. These conditions included the development of footprints by standing or walking astronauts, scooping, sampling, or penetrating the soil with shafts and Apollo lunar handtool mockups. Figure 4–10(a) shows a simulation test. Figure

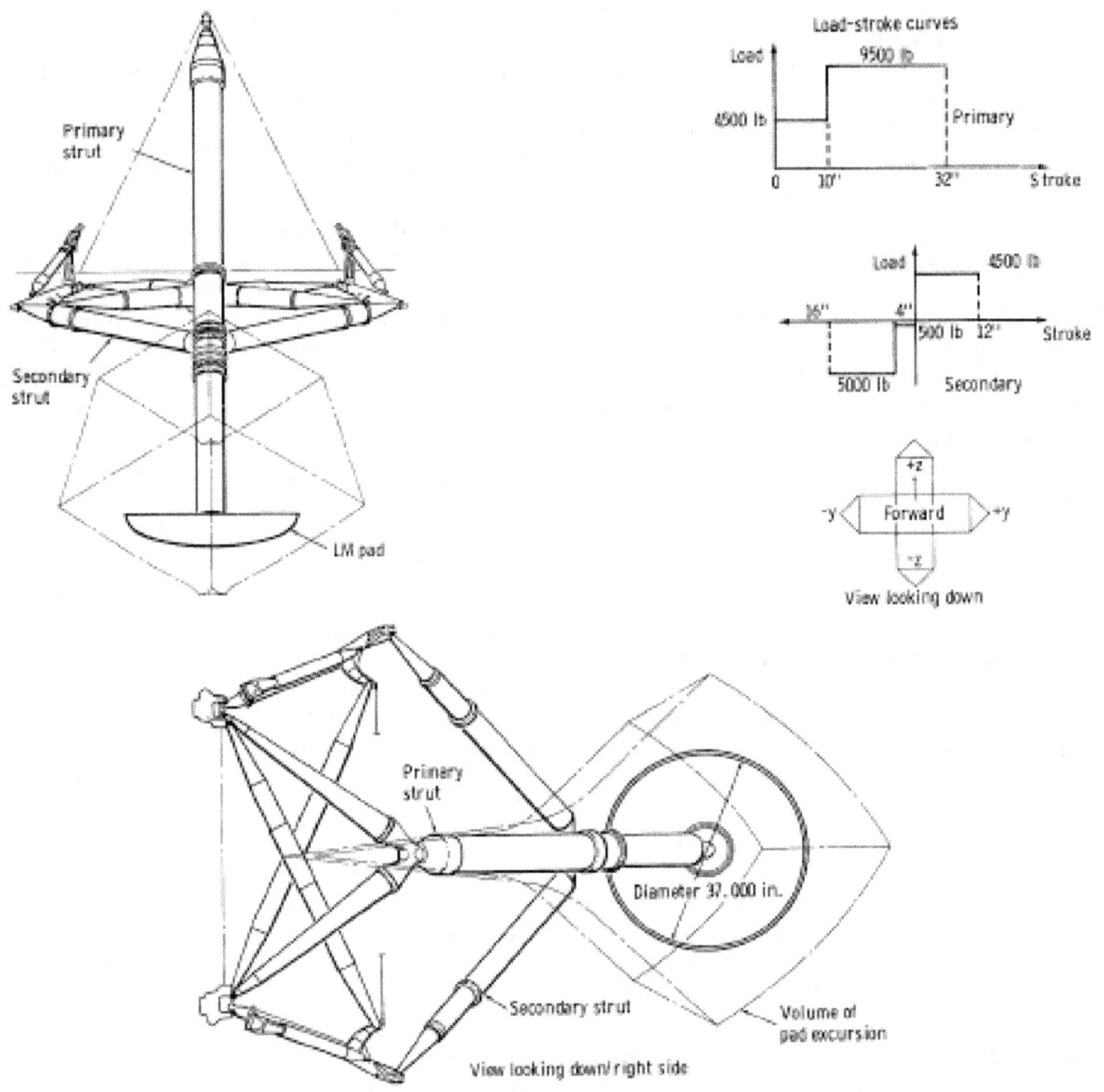

FIGURE 4-1.—Lunar module footpad geometry and load-stroke curves for the primary and secondary struts.

4-10(b) shows astronaut bootprints developed under essentially static loading conditions, and figure 4-10(c) shows astronaut bootprints developed under loading of the same magnitude applied in a manner simulating an astronaut walking. From such tests, inferences were made on the material behavior of the lunar soil during EVA, taking into account the effect of lunar gravity. The results appear to agree reasonably well with the behavior of the material actually observed during the EVA. Based on the current data, postflight terrestrial simulations are planned to gain further insight into the physical characteristics and mechanical behavior of lunar surface materials.

SOIL MECHANICS INVESTIGATION

FIGURE 4-2.—Astronaut Aldrin descending from the LM. Note the configuration of the astronaut boot sole. (NASA AS11-40-5866)

FIGURE 4-3.—Laser Ranging Retro-reflector experiment. (NASA AS11-40-5952)

FIGURE 4-4.—Astronaut Aldrin deploying the PSEP. (NASA AS11-40-5948)

FIGURE 4-5.—Apollo lunar handtools. Pictured from left to right are the hammer, the gnomon, tongs, the extension handle, and a scoop. (LRL photograph S-69-31860)

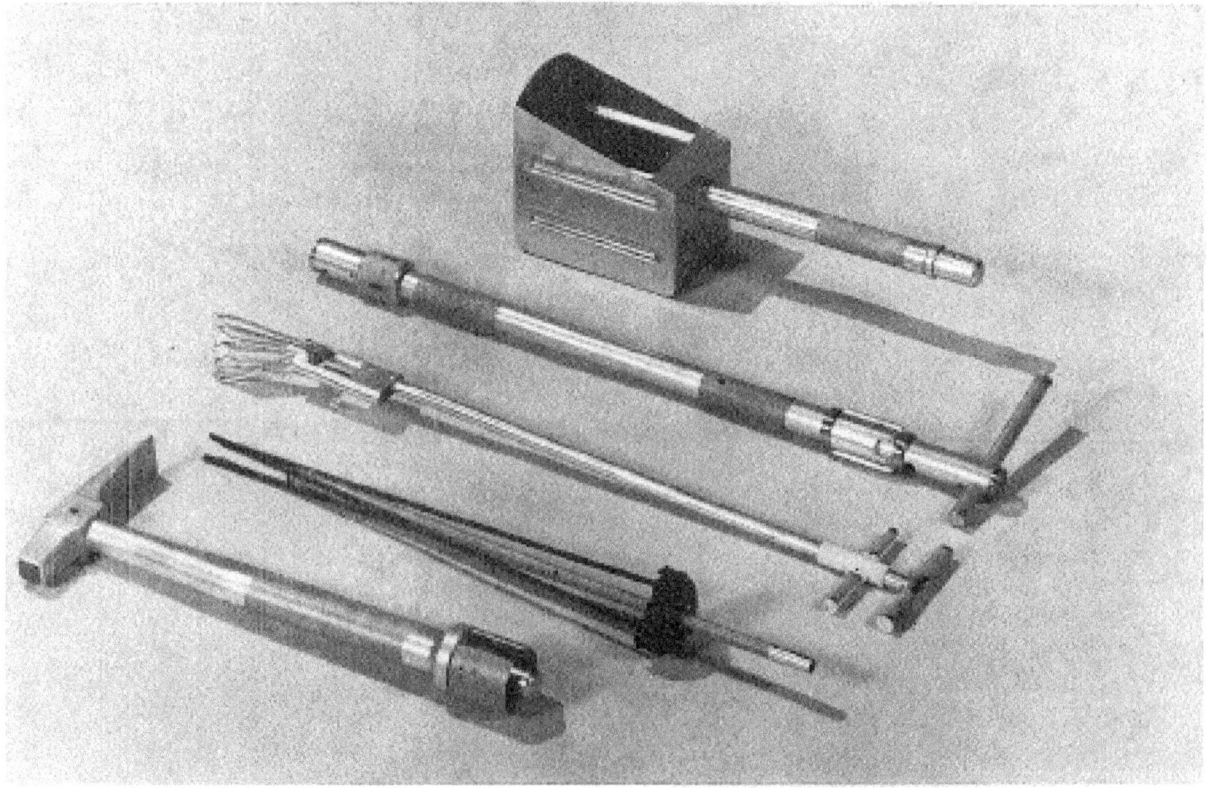

SOIL MECHANICS INVESTIGATION

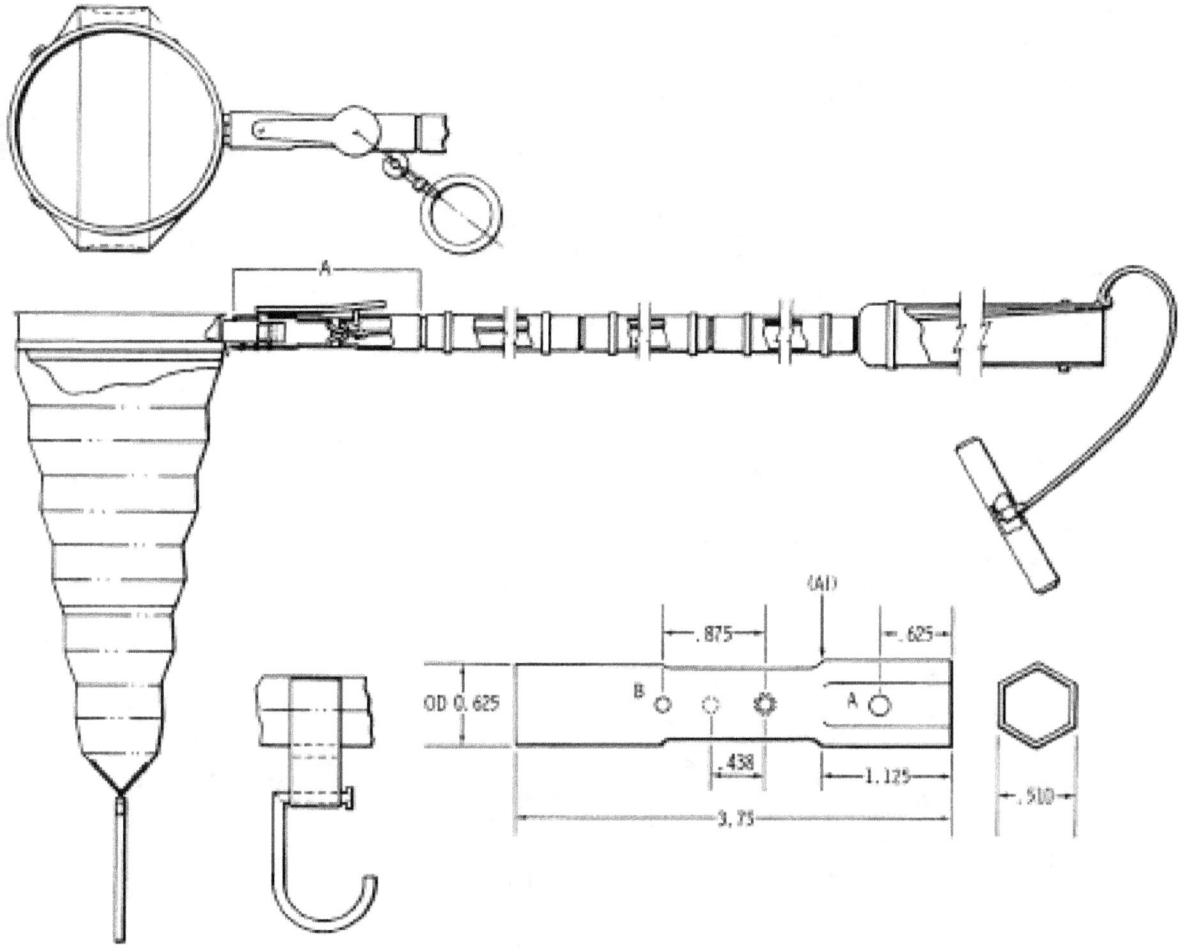

FIGURE 4-6. — Contingency sampler, dimensions are given in inches.

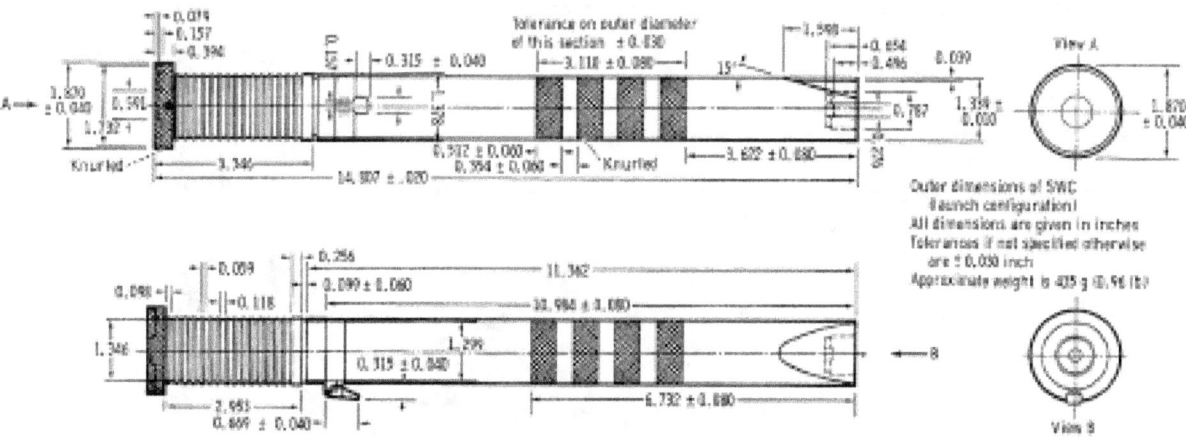

FIGURE 4-7. — Staff of the SWC experiment.

Total length of the flagpole is 92.625 inches.
 (length includes the top of the bracket) and the 0.125 collar
Length of horizontal section (unextended) is 47 inches
Flag dimensions are 36 by 61 inches

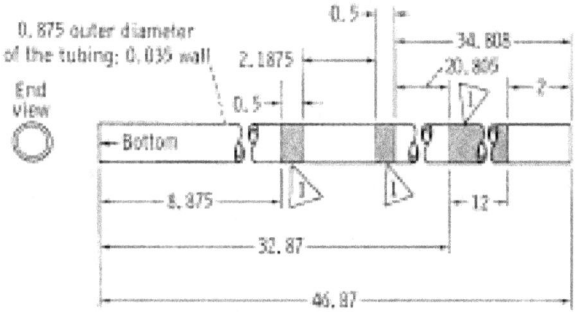

FIGURE 4-8. — The lower section of the flagpole.

Total weight of the flag and the flagpole is 1.0 pound 14.125 ounces

▷ Surface is grit blasted to a "normal" finish, as measured by the Apollo lunar handtool grit blast comparator gage s/n 1001

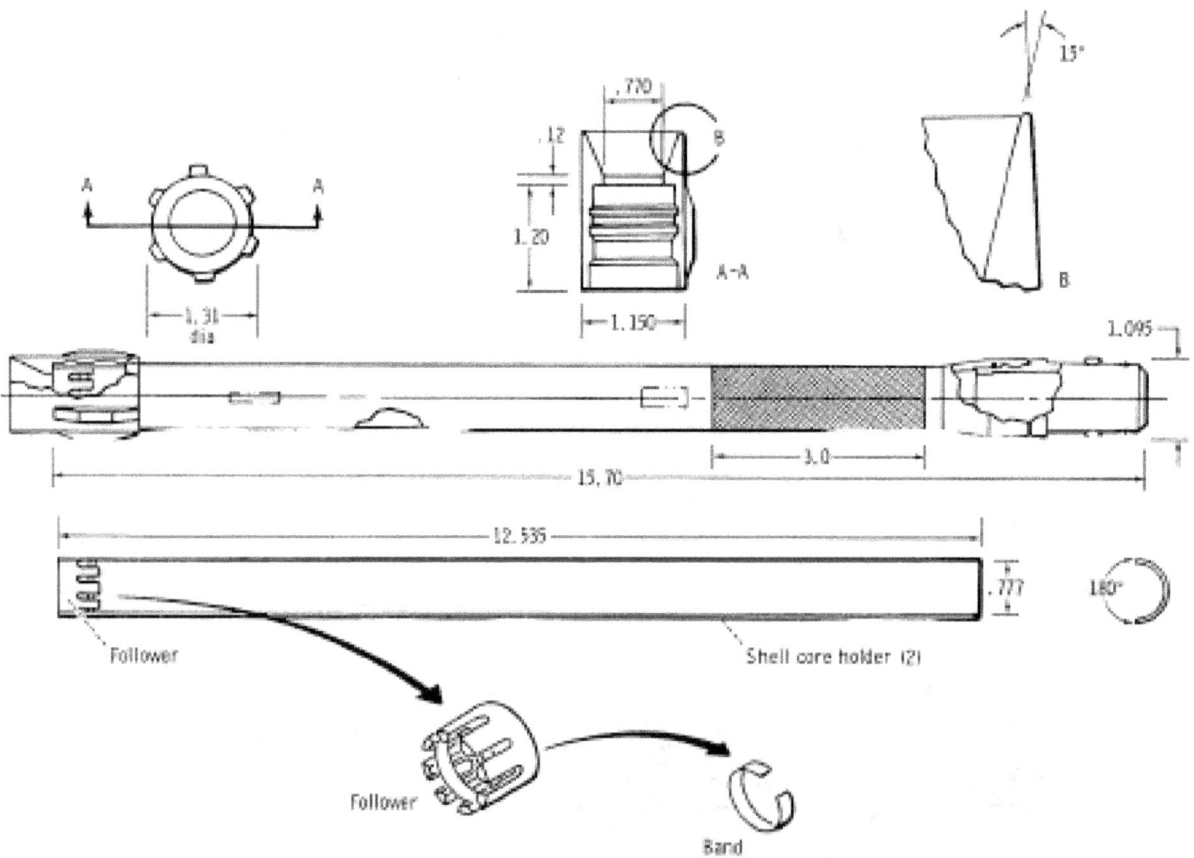

FIGURE 4-9. — Core-tube sampler and bit used on the Apollo 11 mission, dimensions are given in inches.

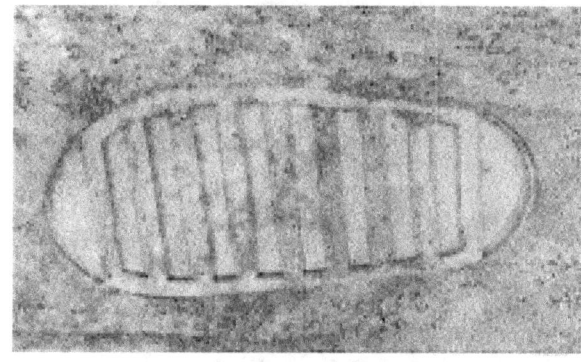
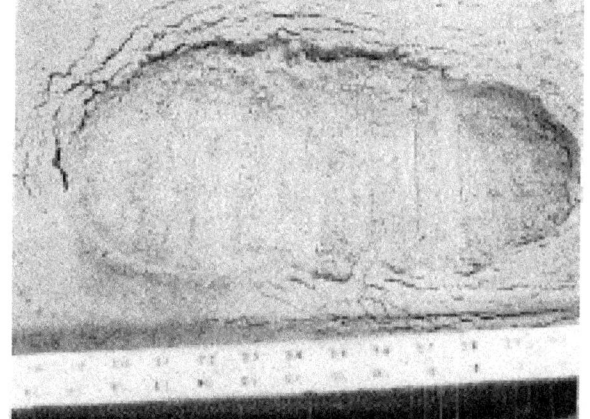

Figure 4-10. — Simulation test. (a) Astronaut bootprint test on simulated lunar soil. (b) Astronaut bootprint on simulated lunar soil developed under action of essentially static load. (c) Astronaut bootprint on simulated lunar soil under action of load of same magnitude as in figure 10(b). Load is applied in a manner simulating a walking astronaut.

Soil Mechanics Data Obtained During the Terminal Stages of LM Descent and Touchdown

Erosion and Visibility

According to discussions of soil properties observed during EVA and obtained from LRL tests, the mechanical behavior of the lunar surface material appears to resemble the behavior of the soil at the Surveyor equatorial landing sites. However, the depth of granular material appears to be variable. The granular material extends from a depth as small as 2 in. over bedrock or rock fragments at some locations to greater depths elsewhere within a radius of 80 to 90 ft around Tranquility Base. The observations of exhaust gas erosion and footpad penetrations must be interpreted in light of this variable soil profile and the motion of the LM during the terminal stages of descent. This discussion will be tied to the astronauts' comments while on the Moon, returning to Earth, and during debriefing.[2] Information was also obtained from spacecraft telemetry, the sequence-camera photographs taken during descent, and the still photographs taken during EVA.

From raw spacecraft telemetry, which was received at 2-sec intervals, the horizontal velocity, the vertical velocity, and the altitude of the LM have been obtained and plotted in figure 4-11. These data may be altered after processing, but the data are thought to be generally indicative of the LM movements immediately before touchdown. The two horizontal velocity components

[2] As taken from the tape-recording transcript, with amendments. Quotations are numbered for easy reference, and quotation numbers are given in parentheses.

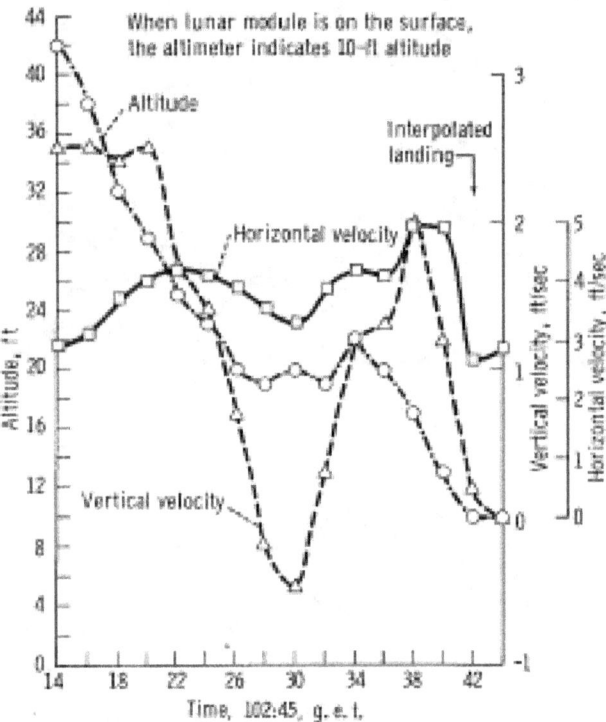

FIGURE 4-11.—Descent of the LM: altitude and velocities versus time.

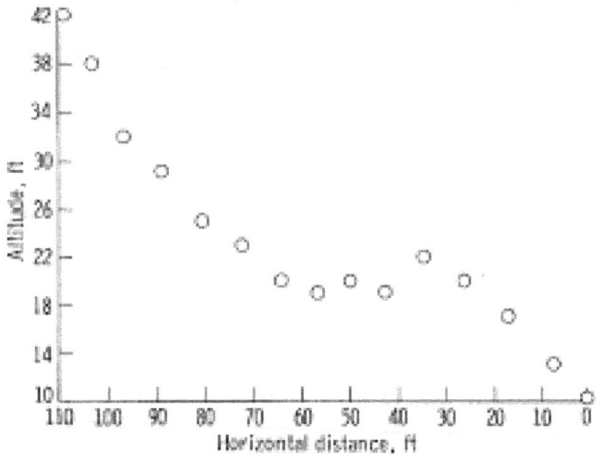

FIGURE 4-12.—Descent of the LM: altitude versus horizontal distance.

are not given separately. This information has been plotted in a different form in figure 4-12 to show the vertical trace of the flightpath. From figures 4-11 and 4-12, the actual touchdown time of the LM footpads, not the LM probes, has been determined to be 102:45:42 g.e.t. (4:17:42 e.d.t.), although selection of the time is complicated by the relatively long interval between data points and by the fact that the telemetry reading made immediately before touchdown is successively repeated after touchdown. However, if 102:45:42 g.e.t. is taken to be the touchdown time, the spacecraft had a relatively low vertical velocity of approximately 1 fps at touchdown and a high lateral velocity between 3 and 4 fps, compared to nominal values of 3 and 0 fps, respectively. From the photographs taken of the footpads and contact probes following touchdown (figs. 4-13(a) and 4-13(b)), it appears that the lateral velocity was to the left of the spacecraft or in the negative y-direction, slightly southeast.

During descent, the following comment (quotation 1) was made: "Forward . . . forward 40 ft, down 2½, picking up some dust." During the scientific debriefing, Astronaut Neil A. Armstrong qualified the in-flight remark by observing that he noticed "substantial haziness" at approximately 100 ft. Astronaut Edwin E. Aldrin, Jr., said he first "saw evidence of disturbed material at about 240 ft." On the Moon (at 110:39 g.e.t.), Astronaut Aldrin made the following observations (quotation 2):

> There's one picture taken at the right rear of the spacecraft looking at the skirt of the descent stage. A slight darkening of the surface color — a rather minimum amount of radiating or etching away or erosion of the surface, even though, on descent, both of us remarked that we could see a large amount of very fine dust particles moving out.

From these remarks, the first observations of surface erosion appear to have been made when the footpads were at an altitude of approximately 90 to 240 ft above the surface. (The altimeter records 10 ft when the spacecraft is on the surface.) From the sequence-camera film, the first visible surface-soil disturbance occurred when the spacecraft was moving across a crater on the flightpath. This crater, which is approximately 200 ft from the LM landing site, is the one mentioned in the comment (quotation 3) which was made by Armstrong at 151:41 g.e.t.:

> . . . I took a stroll back to a crater behind us that was maybe 70 or 80 ft in diameter and 12 or 20 ft deep. . . . It had rocks in the bottom. We were essentially showing no bedrock at least in the walls of the crater at that depth.

SOIL MECHANICS INVESTIGATION

FIGURE 4-13.—Footpads of the LM. (a) The +Y footpad (NASA AS11-40-5920). (b) The −Z footpad (NASA AS11-40-5926).

(a)

(b)

The erosion would, of course, be affected by the surface geometry of the crater. However, from the telemetry data, the altitude of the LM above the crater was approximately 100 ft. Quotation 1 presumably indicates a time at which erosion was strongly developed. The height at which erosion first became noticeable is greater than the predicted height (25 ft) at which erosion should become noticeable, if based on the lunar soil model. Accordingly it is possible that the material being moved was a surface layer possessing less cohesion than the underlying layer (and less cohesion than the lunar soil model size), unless the initial erosion was strongly affected by the crater geometry. On the other hand, the height at which erosion became noticeable confirms that the surface material possesses some cohesion because a cohesionless soil of the same size range as the lunar soil would be moved by the LM exhaust gas at a much higher elevation. (This greater elevation would be approximately 1000 ft, although the astronauts would probably not recognize surface effects at 1000 ft even if erosion were occurring.)

From figures 4-11 and 4-12 (and the 16-mm sequence film), descent appears to have stopped for approximately 10 sec ("flare?") when the footpads were 9 or 10 ft off the lunar surface. The appearance of the ground adjacent to the LM at this descent stoppage is shown in figures 4-14(a) and 4-14(b). As shown in these figures, soil transport by the LM DPS engine exhaust is fully developed, and except for the interruptions in the flow caused by rocks, the surface is obscured. During the postflight debriefing, Astronaut Armstrong reported (quotation 4) that the appearance of the surface was "very much like looking through a thin ground fog. The visibility is probably degraded about 75 percent." Armstrong also noted some difficulty in obtaining a visual reference for lateral control of the spacecraft motion because of the high velocity at which the particles were moving. Consequently, some of the comments made by the astronauts indicate that the blowing dust was "transparent," and other comments give evidence of obscuration. From the sequence-camera film, the lunar surface seems to be concealed by the blowing dust sheet for the final 20 to 30 sec of descent.

(a)

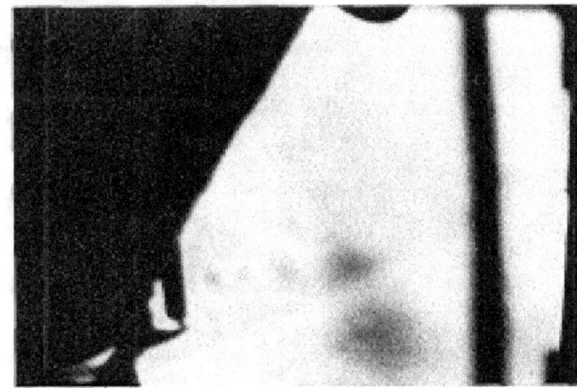

(b)

FIGURE 4-14. — Sequence-camera film frames. (a) Frame from sequence-camera film showing initiation of erosion during LM descent. (b) Frame from sequence-camera film showing fully developed erosion during LM descent.

The distance from the ground zero under the DPS engine nozzle to the farthest point visible in figure 14(a) is several hundred feet. Therefore, probably erosion effects and certainly the eroded soil extended to much greater distances. The following is a subjective impression of this erosion that was given by Astronaut Aldrin at 151:25 g.e.t. (quotation 5):

> I took a number of samples of rocks off the surface and several that were just subsurface and about 15 to 20 ft north of the LM and then I recalled that that area had been probably swept pretty well by the exhaust of the descent engine.

In the scientific debriefing, the astronauts observed that the eroded material traveled a great

Figure 4-15. — Lunar surface under the DPS engine exhaust nozzle. (NASA AS11-40-5921)

distance, even to the point of obscuring their horizon.

The relatively high lateral velocity of the spacecraft in the few seconds before touchdown means that erosion was never as fully developed at one place as would occur under the nominal vertical descent conditions. Concerning this lateral velocity, the astronauts reported the following:

> ALDRIN (quotation 6): "I wonder if that right under the engine is where the probe might have hit."
>
> ALDRIN (quotation 7): "Neil, look at the minus [y?] strut, the direction of travel there, traveling from right to left."
>
> ALDRIN (quotation 8): "This one over here underneath the ascent [descent?] engine where the probe first hit."

However, some surface effects were visible after touchdown as is evident by the following remarks:

> ARMSTRONG (quotation 9): "The descent engine did not leave a crater of any size. There's about 1-ft clearance on the ground. We're essentially on a very level place here. I can see some evidence of rays emanating from the descent engine, but very insignificant amount."
>
> ALDRIN (quotation 10): "There's no crater there at all from the engine."
>
> ARMSTRONG (at 110:39 g.e.t.) (quotation 11): "There's no evidence of problem underneath the LM due to engine exhaust or drainage of any kind."

The amount of observed erosion seems to have been very small. Quotation 9 mentions that the height of the DPS engine nozzle from the lunar surface was approximately 1 ft. To better estimate the height of the engine nozzle above the lunar surface, a measurement of the shadow length on the surface of spacecraft components (including the rocket engine bell) was made from the photographs taken by the astronauts. From this measurement, the height of the engine nozzle above the lunar surface was estimated to be 18 to 20 in. For the distance which the shock absorbers of the landing legs are estimated to have stroked (discussed later), the distance from the nozzle to a level, rigid lunar surface is considered to be approximately 17 in. Because the footpads penetrated approximately 1 to 3 in. into the lunar surface, the nozzle height above an undisturbed, level lunar surface would be 14 to 16 in. An observed height of approximately 1 ft would indicate that the lunar surface below the nozzle was 2 to 4 in. higher than a level lunar surface. However, the estimated height of 18 to 20 in. can be interpreted to mean that 4 to 6 in. of erosion had occurred. The lunar surface may not, however, have been level at this location before the landing. Figure 4-15 shows the surface in the vicinity of the nozzle. From the photograph, it is apparent that some soil must have been removed by the exhaust gas.

Confirmation of this removal might be inferred from the following remark made by Astronaut Armstrong on the small depth of penetration of his boots near the LM footpad (at 109:24:20 g.e.t.) (quotation 12):

> The surface is fine and powdery. I can kick it up loosely with my toe. It does adhere in fine layers like powdered charcoal to the sole and insides of my boots. I only go in a small fraction of an inch — maybe an eighth of an inch, but I can see the footprints of my boots and the treads in the fine sandy particles.

The Surveyor spacecraft landings demonstrated that, although the lunar surface and the underlying lunar soil are dark, the lunar surface is lighter in color than the underlying soil. Thus, any disturbance of the surface is manifested by a dark appearance of the disturbed area. Quotation 2, quotation 9, and the following quotation indicate that the descent engine caused some disturbance.

> CAPCOM (at 114:21 g.e.t., with the astronauts inside the LM after EVA) (quotation 13): "Next topic here relates to the rays which [luminate?] from the DPS engine burning area. We were wondering if the rays [luminating?] from the — beneath the engine are any darker or lighter than the surrounding surface."
>
> TRANQUILITY BASE (quotation 14): "The ones that I saw back in the back end of the spacecraft appeared to be a good bit darker, and of course, viewed from the aft end, well, they did have the Sun shining directly on them. It seemed as though the material had been baked somewhat — and also scattered in a radially outward direction, but in that particular area, this feature didn't extend more than about 2, maybe 3 ft from the skirt of the engine. I wouldn't say it was necessarily material that had been uncovered. I think some of the material might have been baked. . . . Now, in other areas, before we started traveling around out front, why we could see that small erosion had taken place in a radially outward direction, but it had left no significant mark on the surface, other than just having eroded it away. Now it was different right under the skirt itself. It seems as though the surface had been baked in a streak fashion . . . but that didn't extend very far."

Erosion rays appear in figure 4-15. It is also observed in figures 4-15 and 4-16 that the $-Y$ footpad probe touched the surface under the DPS engine (quotations 6 and 8), approximately 10 to 12 ft from the final footpad location. Because the probe is approximately 5 ft long, a general lateral to vertical velocity ratio of 2:1 or 2.5:1 is confirmed. However, a point of interest is that the trench left by the $-Y$ probe near the engine is substantially wider and deeper than is the probe or the mark left by the $-Y$ probe near the footpad. It is concluded that the first disturbance of the soil by the probe, which was followed by engine exhaust impingement on the soil, weakened the soil and permitted more erosion to occur in the vicinity of the first contact of the probe. An indication of exhaust-flow channeling in the trench made by the probe exists. None of the grooves left by the other probes has this appearance; grooves left by other probes are generally of uniform width and uniform depth.

The combination of the lateral component of the LM velocity in the few seconds before touchdown and before the engine shutdown transient probably caused little pressurizing of the soil by the exhaust gas in any specific area. Postshutdown gas-venting effects from the soil would therefore be expected to be minimal. Quotation 11 indicates that the astronauts did not observe any gas-venting effects during EVA, and Astronaut Aldrin made the following comment (quotation 15):

> It was reported beforehand that we would probably see an outgassing from the surface after actual engine shutdown, but, as I recall, I was unable to verify that.

However, in an examination of the sequence-camera film, although there is difficulty in corre-

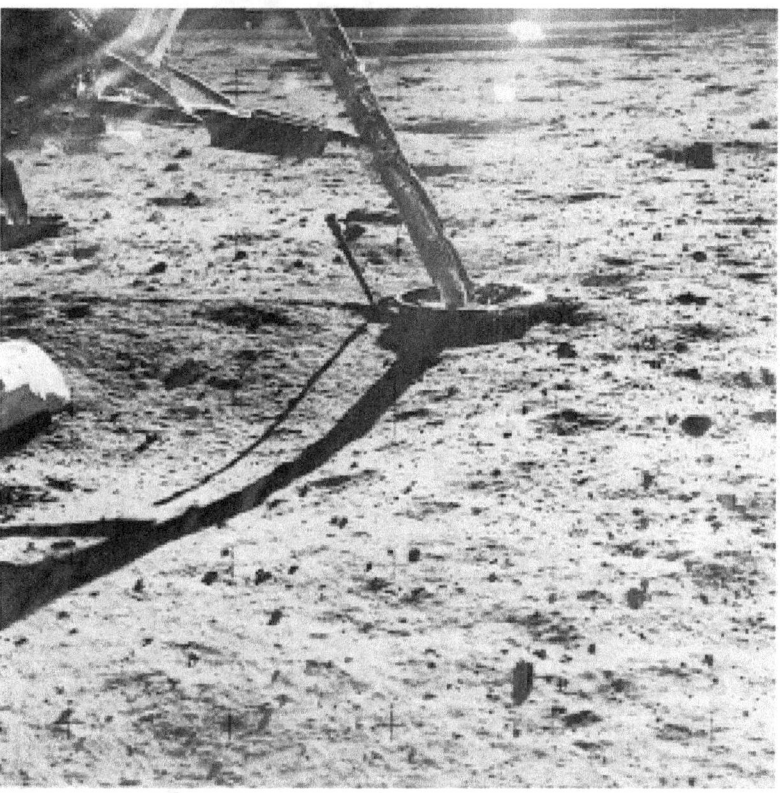

FIGURE 4-16.— The —Y footpad. (NASA AS-11-40-5865)

lating events on the film with spacecraft events, such as engine shutdown, a change in the erosion pattern appears to occur at approximately the same time as engine shutdown. Figures 4–17(a) and 4–17(b), from the last few frames in the film, show this transient effect, which may be caused by either a change in the engine behavior or by an outgassing effect. In figure 4–15 (in the lower right corner and in the bottom of the photograph), numerous fractures appear in the lunar surface in the region that was disturbed by DPS engine exhaust. Because the firing of a rocket engine against a slightly cohesive soil in a vacuum results in a "plucking" type of erosion in

(a)

(b)

FIGURE 4-17.— Two of last few frames from the sequence camera at LM touchdown.

which soil chunks are plucked out of the surface and ejected (ref. 4–14), these fractures could be an indication of incipient erosion of this type of plucking erosion. In addition, venting of the exhaust gases from the soil surface after engine shutdown also causes removal of the soil in chunks or lumps when the soil is cohesive; therefore, the cracks may be evidence of venting. Astronaut Aldrin also remarked after touchdown that he observed some rocks that were fractured, possibly from the DPS engine exhaust.

Because the rocks returned have been found to vary from being moderately hard to fracture to being hard to fracture, it would seem unlikely that any rock fragments were broken by the DPS engine exhaust effects. Aldrin probably observed soil clods or chunks that were removed by the exhaust stream. It is not possible at present to determine the cause of the fracturing.

During ascent from the lunar surface, the LM descent stage is left, allowing the ascent engine exhaust to impinge on the descent stage first. Soil erosion is, therefore, minimized until the ascent stage has reached an altitude that will allow the exhaust to impinge on the lunar surface. Apparently little or no erosion took place during the ascent, as evidenced by the following exchange:

> CAPCOM (quotation 16): "Did you notice — was there any indication of any dust cloud as you lifted off? Over."
>
> ARMSTRONG (quotation 17): "Not very much. There was quite a bit of Kapton[2] and parts of the LM went out in all directions and usually in the same distance as far as I can tell, but I can't remember seeing anything of a dust cloud."

In addition, the solar panels and the dust detector on the PSEP equipment, which were situated approximately 60 ft from the LM, showed no signs of degradation from the lunar soil that was blown away during the ascent.

Touchdown and Penetration

The low vertical velocity and the high lateral velocity of the LM at touchdown resulted in lower penetrations of the footpads into the lunar surface and less stroking of the shock absorbers than would be expected for a nominal landing.

[2]Kapton is an insulating material that was used in the construction of the LM.

With respect to footpad penetration, the astronauts made several comments.

> ARMSTRONG (quotation 18): "I'm at the foot of the ladder. The LM footpads are only depressed in the surface about 1 or 2 in."
>
> ARMSTRONG (at 110:39 g.e.t.) (quotation 19): "I don't note any abnormalities in the LM. The pads seem to be in good shape."
>
> ALDRIN (quotation 20): "It's very surprising, the surprising lack of penetration of all four of the footpads. I'd say if we were to try and determine just how far below the surface they would have penetrated, you'd measure 2 or 3 in., wouldn't you say, Neil?"
>
> ALDRIN (at 110:50 g.e.t.) (quotation 21): "We are back at the −Z strut now. The stereopair we are taking illustrate [the] very little force at impact we actually had."

From these comments, it is apparent that the range of footpad penetrations was 1 to 3 in. into the lunar material that remained after most of the DPS engine erosion of the surface had occurred. This range is confirmed by the depths of penetration visible in figures 4–13(a) and 4–13(b). In these photographs, it also appears that some soil may have been deposited against the footpad-to-ground contact region as a result of rocket engine erosion after touchdown. Alternatively, the soil that was plowed up by the footpad in its lateral motion may have been eroded by the DPS engine exhaust after touchdown.

The descent ladder on the LM is attached to the fixed portion of the landing gear, and the footpad, by compressing the shock absorber, can move with respect to the ladder. This movement decreases the distance from the bottom step of the ladder to the footpad. With no compression of the shock absorber, the distance is approximately 3 ft. The astronauts' remarks show that hardly any stroking of either the primary or the secondary shock-absorbing struts occurred.

> ARMSTRONG (at 109:20 g.e.t.) (quotation 22): It's not even collapsed too far, but it's adequate to get back up . . . it takes a pretty good little jump."
>
> ARMSTRONG (quotation 23): "That's a good step. About a 3 footer."
>
> CAPCOM (quotation 24): "Can you estimate the stroke of the primary and secondary struts?"
>
> TRANQUILITY BASE (quotation 25): "All the

Figure 4-18. — Primary strut and secondary strut of the +Y footpad after touchdown. (NASA AS11-40-5919)

struts are about equally stroked and the height from the ground to the first step is about 3 ft or maybe 3½ ft."

From figure 4-18, the stroking of the primary shock can be estimated to be approximately 0 to 1 in.

It is apparent, therefore, that the astronauts achieved an almost static landing on the lunar surface with regard to the landing gear. For the vertical load of approximately 700 lb on each LM footpad acting statically on a circular loaded area resting on a Surveyor soil model several feet deep, the radius of the contact area would be approximately 9 in. With this contact area, the required penetration of the spherical footpad into a lunar surface consisting of the lunar soil model material previously described is approximately 2.5 to 3 in., which is close to the maximum penetration that was observed. Lesser penetrations may be because of the presence of rock fragments or stronger material below the surface.

Soil Mechanics Observations During the EVA

> The surface is fine and powdery. I can kick it up loosely with my toe. It does adhere in fine layers like powdered charcoal to the sole and insides of my boots. I only go in a small fraction of an inch — maybe an eighth of an inch, but I can see the footprints of my boots and the treads in the fine sandy particles.

These observations, made by Astronaut Armstrong immediately after his first step onto the surface of the Moon, provided the first "ground truth" data against which previous conclusions concerning the mechanical properties of lunar surface materials could be examined. During the 2½ hr of EVA, additional information became available that can be extracted from the descriptions and comments of the astronauts, television photography, still-camera and sequence-camera photography, returned samples, and postflight simulations.

A more detailed account of the lunar surface characteristics as deduced from lunar surface

EVA is given in the following sections of this report.

General Characteristics of the Landing Site

The LM landed in a field of small, elongate, and circular craters located at 0.67° N and 23.49° E; a point approximately 130 m north and 600 m west of West Crater, a relatively fresh, sharp-rimmed ray crater that is approximately 200 m in diameter. The crater is surrounded by ejecta containing blocks ranging to more than 3 m in size and extending 100 to 200 m from the crater rim. Several hundred feet to the right of the LM, in a northerly direction, there is a boulder field with boulders a meter or larger in size. A hill was estimated to be from ½ to 1 mile west of the LM.

The LM attitude on the surface was tilted to the east 4.5° from the vertical and yawed left (to the south) approximately 13°. The immediate region around the LM was relatively free of rocks and was covered with craters varying from approximately 100 ft to less than 1 ft in diameter. Figure 4-19 is a plan view of the landing site indicating major crater locations, other topographical features, and astronaut activities. Figure 4-20 is a photographic mosaic showing a panoramic view of the landing site.

Although photographs show the surface at the landing site to be relatively smooth, the real-time impression of the crewmen was that the site was very rough with many slopes, holes, and ridges, and that the general topography consisted of undulations 1 m or larger in amplitude. Further conclusions regarding the lunar terrain roughness characteristics at the landing site await the construction of detailed topographic maps.

Description of the Surface Material

The ground mass at the Apollo 11 LM landing site is a fine-grained, granular material, consisting of bulky grains in the silt to fine-sand particle size range. Larger rock particles, to 4 ft in size and varying in shape and type, were distributed throughout this material. Some of the rocks were lying on top of the surface, others were partially buried, and others were partially exposed with the top part of the rock flat with the surface. The only evidence of possible bedrock was in West Crater. On the basis of observations of craters, the astronauts reported absence of layering in the subsurface material. On the surface, local variations in coarse particle sizes did seem to exist.

The color of the returned fine-grained material was medium-dark-gray (N-3 to N-4 on the Geological Society of America (GSA) color scale). However, on the lunar surface, the color and brightness of the material appeared to change with the Sun angle and with the viewing angle. At higher Sun angles, the color of the material was toward the brown end of the spectrum, whereas at low Sun angles, the color of the material was toward the slate-gray end of the spectrum. The down-Sun region was extremely bright and appeared to be light tan, whereas when the astronauts proceeded back toward the cross-Sun region, the brightness of the surface diminished, the surface color began to fade, and the material began to appear more grayish-cocoa. In the shadow, the lunar surface appeared to be very dark. Although the astronauts were able to see into the shadow, visibility was low. Regions disturbed by the spacecraft landing and the astronaut activities were considerably darker than the undisturbed area, as may be seen in figure 4-21. The degree of darkening that developed as a result of disturbance appears to be a function of the phase angle; i.e., angle between the Sun, the point being observed, and the observer. It has not yet been established conclusively whether this darkening is caused by a change in surface texture; i.e., particle sizes and roughness, as a result of disturbance, or because the material underlying a thin surface layer is actually darker. A similar darkening phenomenon was observed at the Surveyor landing sites and is discussed in detail in the Surveyor reports.

The fine-grained surface material had a powdery appearance and was easily kicked free as the astronauts moved on the surface. The particles moved along ballistic trajectories according to a definite pattern that seemed to depend on the angle of impact of the boot with the lunar surface. A small amount of material landed in a linear pattern, but most of the material landed in regions that were generally radially away from the boots. Kicking at the same angle with different forces did not seem to make too great a

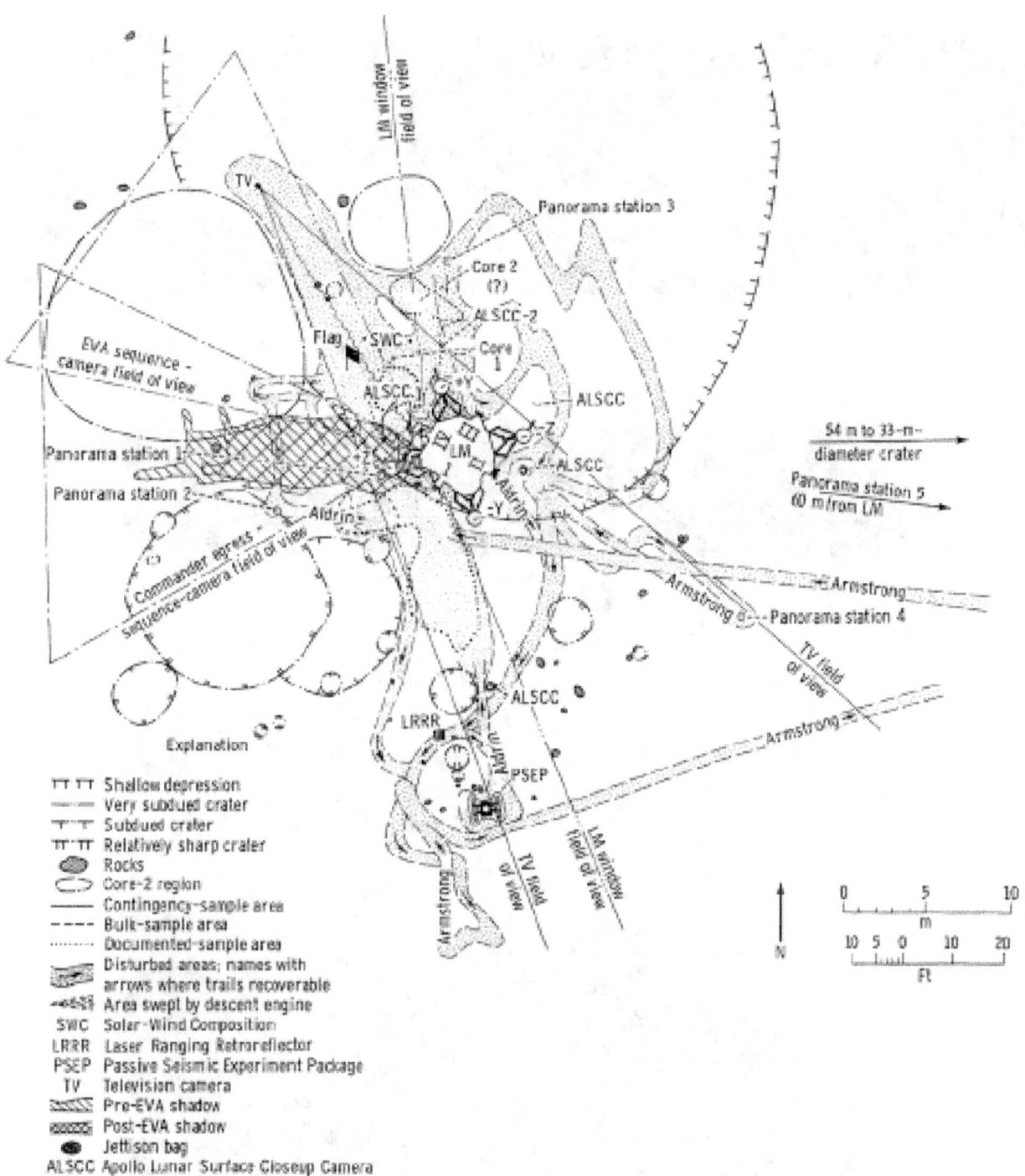

FIGURE 4-19.—Preliminary traverse map of the Apollo 11 LM landing site.

difference in extending the pattern. The distribution of particles that struck the surface appeared to be uniform, and, apparently, the material did not move in its finest particulate size, but tended to form clods approximately 5 to 10 mm in size. As expected, and as was the situation

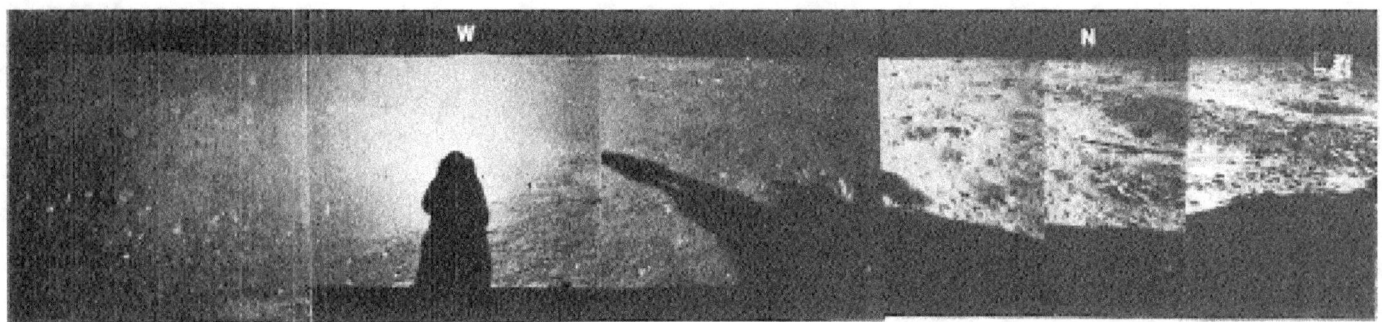

FIGURE 4-20. — Photo mosaic showing the panorama of the landing site.

FIGURE 4-21. — Region in the vicinity of Tranquility Base showing the darker region that has been disturbed by astronaut activities and the lighter colored, undisturbed region. (NASA AS11-37-5516)

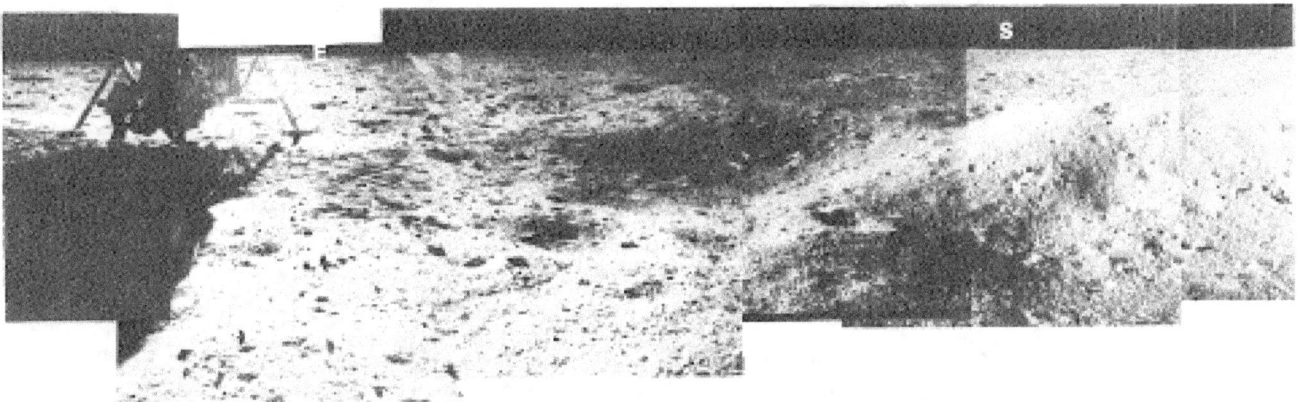

with the surface soil erosion caused by the DPS engine exhaust, kicking of the surface material did not leave a residual, lingering dust cloud to settle slowly to the surface.

Adhesive Characteristics

The loose, powdery, fine-grained surface material exhibited adhesive characteristics, as a result of which the material tended to stick to any object with which it came into contact, including the astronauts' boots and spacesuits, the television cable, the Lunar Equipment Conveyor (LEC), and the corners of the solar panels of the PSEP that came into contact with the surface during deployment of the seismometer. The largest particles that adhered to the spacesuit were of fine-sand size. The majority of the particles fell off the spacesuits when the LM strut was kicked by the astronauts. However, discoloration of the fabric meant that some of the particles must have continued to adhere. Figure 4-22 shows Astronaut Aldrin and the substantial discoloration of his boots and spacesuit.

Adherence of the fine, powdery material to the television cable and the LEC imposed some operational problems. First, when the television cable was coated with this dust, the cable blended with the lunar terrain and was not easily discerned by Armstrong who repeatedly became entangled in the cable. Second, as the LEC was operated, the powdery material adhering to it was carried into the spacecraft cabin, posing a housekeeping problem. As the LEC went through the pulleys, sufficient fine-grained material collected in the pulleys so that they

FIGURE 4-22. — Astronaut Aldrin showing boots and suit discoloration caused by adhering lunar soil. (Enlargement from NASA AS11-40-5903)

tended to bind. Armstrong commented, "You can actually feel the gritty material getting caught up in the rollers of the pulley." Then the material that was shaken off came out on the downward side of the pulley, and Armstrong further remarked:

> The part of the LEC that was coming down would rain powder on top of me, the MESA (Modularized Equipment Stowage Assembly), and the SRC's (Sample Return Containers) so that we all looked like chimney sweeps.

The fine, powdery material also adhered to lunar rock samples that were brought back to Earth and left a trace of fine dust that coated the core tubes that were returned to the LRL. This material adhesion, however, was not of sufficient magnitude to offer any resistance to pulling of objects inserted into the lunar surface, such as the SWC experiment staff, the flagpole, or the core tubes.

Slipping of Dust-Covered Objects in Contact With Each Other

The thin layer of material that adhered to the soles of the astronauts' boots produced a tendency to slip on the LM ladder after EVA. Similarly, the powdery coating of the rocks on the lunar surface was observed to make the rocks slippery. Study of particle shapes in returned lunar samples has indicated a substantial proportion of small (± 50 μm) spherical particles. The suggestion has been made that these small spherical particles might behave in a manner similar to ball bearings, giving rise to slipping tendencies. The observed behavior is not unusual, however; a fine terrestrial dust, when confined between two relatively hard surfaces, such as the astronaut boot and the spacecraft ladder rung or the astronaut boot and a rock surface, will also produce the same tendency to slip. This slipping tendency disappears when a relatively solid body is moving in a large mass of soil, as was the general case with the crew who experienced no serious mobility problems in walking, jumping, or loping; and reported that adequate traction existed for starting, turning, and stopping. The fact that the soil may contain small spherical particles is relatively unimportant with regard to this slipping tendency.

Deformation and Strength Characteristics

The nature of the data provided by the LM landing and EVA dictated that analyses of the strength and the deformation characteristics of the lunar soil in the vicinity of Tranquility Base must be primarily descriptive and qualitative rather than quantitative. This is because no force or deformation measuring devices were used during the lunar surface activities. In general, however, the soil behavior appears to have been consistent with the behavior that would be expected for a soil having properties characteristic of the soils studied at the Surveyor equatorial landing sites. These properties were previously summarized in this report.

Cohesion

The tendency of the fine-grained material to form clods indicates that the material possesses a small, but finite, amount of cohesion. In some situations, soil clumps could not be distinguished from rock fragments; however, when these soil clumps were stepped on, they tended to crumble. According to the astronauts, such clods of fine-grained material constituted about 10 percent of the blocky material that protruded above the surface. Other evidence of the slightly cohesive nature of the lunar soil was given by the following observations:

(1) Initially loose and fluffy material readily compacted under a load and retained the detail of a deformed shape, as seen in figures 4–23 to 4–25. As a result, astronaut mobility on previously traversed surfaces was enhanced.

(2) The material could stand unsupported at a height of at least a few inches on vertical slopes.

(3) During the bulk-sample collection, it was observed that as the scoop cut through the lunar soil, the remaining material left a sharp, solid edge. The material that went into the scoop, however, crumbled with no evidence of aggregation of the particles.

(4) The holes made by the core tubes appeared to remain intact upon the removal of the tubes.

(5) When the core tubes were rotated to an inverted position to keep the core sample from coming out, there was no tendency for the material to pour out. According to Aldrin when the core-tube bit was removed, "The surface seemed to separate again without any tendency for the material to flow or move." The material

SOIL MECHANICS INVESTIGATION

FIGURE 4-23.—Undisturbed lunar surface. (NASA AS11-40-5876)

FIGURE 4-24.—Same surface as shown in figure 4-23 after having been stepped on by an astronaut. Note the bulged and cracked region in front and to the sides of the footprint. (NASA AS11-40-5877)

FIGURE 4-25.—Typical astronaut footprints on the lunar surface in the vicinity of the flagpole. (Enlargement from NASA AS11-40-5874)

collected at the bottom of the core tube had an appearance similar to that of a moist terrestrial soil. According to Aldrin:

> It was adhering or had the cohesive property that wet sand would have. Once it was separated from the cutter, there was no tendency at all for it to flake or to flow.

Frictional Characteristics

Although the surface layer behaved as if it were loose and readily deformable when unconfined, confinement of the soil led to a significant increase in resistance to deformation — a characteristic of soils deriving a large portion of their strength from interparticle friction. Relevant material properties can be assessed from the observations in the following paragraphs.

(1) Relatively small footpad penetrations of 1 to 3 in. were observed. As noted previously, the landing was essentially static. For penetrations in the 1- to 3-in. range, the static bearing pressures exerted by the LM on the lunar surface would be in the range of 2.1 to 0.8 psi. For a Surveyor soil model with a density of 1.5 g/cc, a cohesion of 0.1 psi, and a friction angle of 35°, the bearing capacity would be 9.1 psi for the LM footpad contact area, corresponding to a 1-in. penetration, and 11.5 psi for the LM footpad contact area, corresponding to a 3-in. penetration. These values are based on the following equation for ultimate bearing capacity q_{ult} on level ground, according to the Terzaghi theory for incompressible deformation under a strip footing adjusted for a circular loaded area:

$$q_{ult} = 1.2cN_c + 0.4B\gamma N_\gamma + 1.2D\gamma N_q \qquad (4\text{-}1)$$

where γ is the unit weight of the soil, B is the diameter of the loaded area, c is the unit cohesion of the soil, D is the depth of the footing, and N_c, N_γ, and N_q are bearing capacity factors that depend on the angle of internal friction ϕ of the soil.

On this basis, the pressures applied by the LM footpads are seen to be several times less than the probable ultimate bearing capacity, and, therefore, the small observed values of footpad sinkage would be expected. Because of the small magnitude of lunar gravity, the relatively small width of the loaded area, and the shallow depth below the surface at which the load is applied, the first term of equation (4-1) accounts for the greatest portion of the bearing capacity. Thus the cohesion, in spite of its small value, is a key characteristic of the lunar soil with regard to the ability of the soil to support bearing loads. In the analysis discussed previously, if a unit cohesion of 0.05 psi had been assumed rather than 0.1 psi, the corresponding bearing capacities would have been reduced to 6.2 and 8.6 psi.

(2) The depth of astronaut footprints was relatively small. Typical footprints shown in figure 4-25 correspond to sinkages of the order of one-half inch. From the known weight of a space-suited astronaut, approximately 64 lb in lunar gravity, and one boot sole area, approximately 65 in.², the static bearing pressure when the astronaut weight was applied on one boot was approximately 1 psi. From the known boot dimensions, 13 in. in length and an average width of 5⅛ in., the boot may be assumed to behave as a strip footing. Using the soil model previously discussed, bearing capacities of 3.8 psi and 6.3 psi are computed for unit cohesions of 0.05 psi and 0.10 psi, respectively. Again, the applied stresses are well below the ultimate bearing capacity. It is important to recognize that while the first term, the cohesion term, of the bearing capacity equation (4-1) dominates, the value of N_c increases almost exponentially with increases in the friction angle. Thus, only because the lunar soil is dominantly frictional in nature can the cohesion term assume larger values (relatively) for the small unit cohesion.

(3) The bearing strength of the material supporting the PSEP (fig. 4-4) appeared to increase with increased confinement. Both the PSEP and the Laser Ranging Retroreflector (LRRR) experiment were placed on essentially soft material. Thus, the crewmen could "jiggle the packages down and get them reasonably set into the sand." However, once the PSEP pallet was firmly set into the lunar soil, the astronauts experienced considerable difficulty in "getting it to sink down a little more on one side," in an attempt to change the slope of the PSEP.

(4) The soft spots encountered during EVA were observed by both astronauts to be located at the rims of fresh craters only a few feet in diameter that consisted of loose, very fine-grained material with essentially no large rock fragments

Figure 4-26. — Astronaut footprints at the top edge of a soft-rimmed crater. (NASA AS11-40-5946)

present. The astronauts commented that if a man stepped on the ejecta of such craters away from the crater rim, his foot might sink only a few inches. However, close to the rim (especially on the upper edge of the inside slope), where the material is loose and relatively unconfined, the foot might sink as much as 6 to 8 in. Astronaut footprints left at the soft rim of such a crater can be seen in figure 4-26.

Close examination of this photograph (fig. 4-26) suggests that failure was incipient on the crater slope adjacent to the deep footprint (closest to the crater edge in the foreground). The combined weight of the astronaut and the EASEP package, which he was carrying while walking by the crater rim, produced a unit bearing pressure of 1.4 psi on one boot area in the lunar gravity field. By using Meyerhof's bearing capacity factors (ref. 4-15) for foundations on slopes, the relationships shown in figure 4-27 may be developed for different values of slope angle, friction angle, and cohesion, and for the boot dimensions given previously.

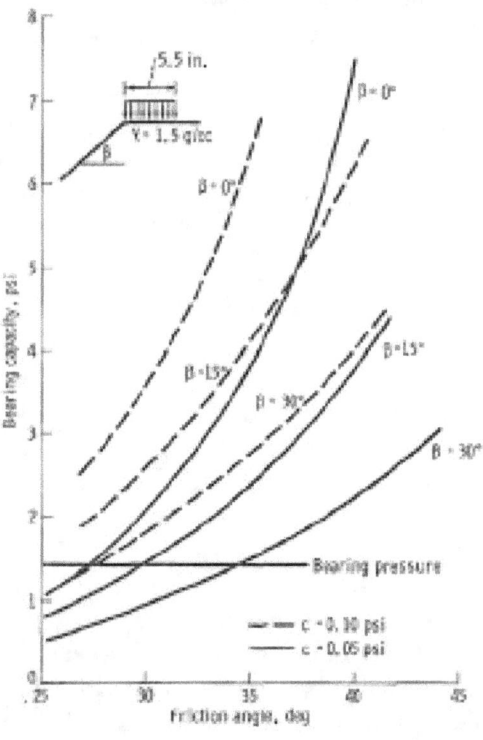

Figure 4-27. — Bearing capacities under astronaut boot adjacent to slopes, for various soil conditions.

Although the exact slope angle of the crater wall has not yet been determined, preliminary estimates indicate the slope angle to be approximately 30°. From figure 4-27, it may be seen that for a 30° slope angle, a friction angle of 35°, and a cohesion of 0.05 psi, failure would be incipient. If the unit cohesion were 0.10 psi, the bearing capacity should be higher. It should be noted, however, that the computations have been based on failure in general shear (incompressible deformation). In fact, figure 4-27 shows that much of the boot sinkage involves compression; thus, the estimated bearing capacities are high. It appears, therefore, the assumed soil properties account reasonably well for the observed behavior.

Deformation Characteristics

Although many of the footprints appear to have resulted from soil compression, in one instance footprints 2 to 3 in. deep were reported to be accompanied by bulging of the surrounding surface and by the formation of cracks extending to distances of several inches. An example of this type of soil deformation can be seen in figure 4-24. Such deformation behavior reflects the presence of cohesion and deformation dominated by shear rather than by compression. Such deformation behavior is consistent with the behavior that was observed in some of the bearing tests conducted by the surface sampler during the Surveyor 3 and Surveyor 7 missions. In general, the deformation behavior of the lunar surface soil appears to involve both shear and compression.

Subsurface Conditions

Several observations made by the astronauts during the EVA indicate that the soft surface material extends only to a depth of a few inches throughout the Tranquility Base region, as evidenced by either static or dynamic penetration resistance. Pertinent observations included the following:

(1) A hard surface was encountered at depths of 1 to 3 in. during contingency sampling. Because the contingency sample was taken from a point close to the LM (i.e., in a location which probably had been eroded by the DPS engine exhaust) the original thickness of soft material may have been greater than the observed thickness.

(2) The bulk sample could be scooped readily only to depths of 1 to 3 in. This sample also was taken from a region close to the LM.

(3) The flagpole (fig. 4-28) penetrated approximately 7 in. into the soil. This penetration was estimated from the distance of the knurled markings on the pole above the surface. The first 4 or 5 in. were easy to penetrate. After these first 4 or 5 in., the resistance to penetration by the underlying soil increased. However, the lunar soil apparently did not provide high frictional support along the shaft because upon ingress into the LM, the astronauts noticed that the entire unit had rotated about the flagpole axis and that the flag no longer pointed in the same direction that it did when originally installed. Whether the flagpole actually tipped slightly is not clear from the available information. Preliminary analysis of the possibility of failure similar to that for a laterally loaded pile indicates that for Surveyor model soil properties, such a failure should not develop under the magnitude of lateral loading induced by the flag.

(4) The SWC experiment staff (fig. 4-29) penetrated the surface to a depth of only 5 to 5½ in., as can be estimated from the knurled markings of the staff that were visible above the surface. The staff was reported not to be as stable as desired, indicating that while resistance to vertical penetration was high, adhesion of the soil along the shaft was low.

(5) Both core tubes driven into the soil in the immediate vicinity of the SWC (fig. 4-29) were easily pushed into the surface to depths of 3 to 5 in. Further penetration of only approximately 2 in. was accomplished, and this further penetration required that the astronaut hammer as vigorously as possible — to the degree that the hammer dented the extension handle attached to the core tube. The core tubes were also reported to remain loose in the soil. They did not remain in place and could be pulled from the soil without resistance. The current core-tube design did not allow for the return of the driving bit. Examination of the bit and its contents after driving would have provided valuable information on the nature of the deeper subsurface material.

SOIL MECHANICS INVESTIGATION

FIGURE 4-28. — Flagpole penetration of the lunar surface. (NASA AS11-40-5905)

FIGURE 4-29. — Astronaut driving core tube near the deployed SWC experiment.

(6) Only one observation suggested that the soft layer in the immediate vicinity of the LM extended to a depth of more than 6 in. Following collection of the contingency sample, Armstrong commented that he could push the sampler handle approximately 6 to 8 in. or more into the surface. It is not clear, however, whether this was indicative of a deep soft layer. Photographs obtained with the sequence camera indicate that the sample holder was pushed into the ground in the same region where the contingency sample had been collected. Thus, the sampler holder may have penetrated soil that had been previously loosened by sampling activities.

Discussion

Whether the high resistance to penetration was due to bedrock, heavily consolidated soil, or a layer of rubble cannot be determined from the available data. The astronauts made the following comments:

> Various tools were pushed into the ground to various depths on numerous occasions. Experience was always the same on this wide variety. On some occasions, rock would be hit and it was clearly evident when that occurred. When buried rocks were hit, it was known that a rock had been hit. In most other instances, it was quite different. The tool would just go in, the force would keep getting higher and higher, and suddenly it just wouldn't go further.

In discussing the resistance exhibited by the subsurface material to core-tube driving, both astronauts excluded the possibility of the core tubes hitting a rock layer or rock fragments. Under the impact of successive blows of the same intensity on the core tube, the corresponding core-tube penetrations appeared to decrease exponentially. Also, no discontinuity appeared to exist in the material hardness. The material was more compact and more cohesive as the tube was driven in deeper.

Data obtained with the surface sampler during the Surveyor 3 and Surveyor 7 missions indicate that the density and strength of the lunar soil increases with depth. This fact and the fact that ultimate bearing capacity increases with depth mean that resistance to penetration should increase with depth of penetration.

From the results of tests on a lunar soil simulant (ground basalt) reported by Mitchell et al. (final report on NASA contract NAS 8–21432), it was deduced that the resistance to penetration by flat-ended rods pushed into the actual lunar soil should increase by approximately 3 psi for each inch of penetration if the bulk density of the soil was the same as that of the core-tube lunar soil samples returned to Earth. (See following sections.)

Data from simulations by space-suited astronauts under reduced gravity conditions have established that an astronaut should be able to exert a downward force of approximately 20 lb on the lunar surface. By using these data and by assuming no significant increase in the strength or density of the lunar soil, table 4–I has been prepared to compare values of observed lunar surface penetrations with predicted values. In the computations, it has been assumed that the penetrators, although hollow, became plugged at a shallow depth of penetration and subsequently behaved as flat-ended penetrators.

From the results listed in table 4–I, it appears that the resistance to penetration of the lunar soil can be accounted for in each instance without postulating the presence of an overconsolidated layer, a cemented layer, or bedrock. It does not follow, however, that such a layer did not exist at any or all of the locations. This could

TABLE 4–I. *Comparison of observed and predicted lunar surface penetrations*

Penetrator	Diameter, in.	Area, in.²	Depth of penetration, in.	Predicted penetration to 20-lb resistance, in.
Contingency sampler handle	0.625	0.31	>6 to 8	21.5
Flagpole	0.875	0.60	7	11
Core tube	1.31	1.35	3 to 5	5
SWC staff	1.34	1.41	5 to 5.5	4.75

only be determined by additional subsurface exploration; i.e., by digging down several inches.

The high resistance to driving of the core tubes into the lunar surface to depths greater than approximately 6 in. cannot be accounted for by the previous arguments because core-tube driving involves dynamic penetration. It may be noted, however, that the hammer used to drive the core tube had a mass of only 827.4 g. In addition, Astronaut Aldrin reported that he was unable to strike the core tube as hard as he would have liked because of the need to steady the tube with one hand, resulting in a restraint on his freedom of movement.

Astronaut and Lunar-Roving-Vehicle Mobility Considerations

As expected, the soil conditions at the landing area did not pose serious restraints on astronaut mobility. The lunar surface provided adequate bearing strength for standing, walking, loping, and jumping; and sufficient traction for starting, turning, and stopping. The ⅙-g lunar gravity conditions provided a pleasant environment in which to work and to maneuver objects. The astronauts quickly adapted to it. As Aldrin commented,

> When your feet are on the surface, you can do fairly vigorous sideways movements such as leaning and swinging your arms without a tendency to bounce yourself up off the surface and lose your traction.

Walking on the surface tended to compact the loose, virgin material. Accordingly, crew mobility was enhanced in previously walked-on regions. Also, from the appearance of the footprints left on the surface, the boot ribs appeared to increase traction.

Confidence in walking on virgin areas was lowered by two factors: (1) the tendency to slip when the boot was in contact with dust-covered rocks and (2) the variability in the softness characteristics of the lunar surface with little apparent change in the local surface topography. However, these factors did not seriously hinder EVA or astronaut performance on the lunar surface.

As expected, small, fresh crater walls having slope angles of 15° or less could be readily negotiated by the astronauts, although it appeared that going straight down and straight up the slopes would be preferable to sideways traversing of these slopes. As Aldrin commented, "The footing is not secure because of the varying thickness of unstable material which tends to slide in an unpredictable fashion."

The material on the rims and the walls of larger sized craters, with crater-wall slope angles ranging to approximately 35°, appeared to the astronauts to be more compact and more stable than the material on the craters that were traversed. Both astronauts remarked that if time had permitted, they would have ventured to the bottoms of these craters.

Commenting on potential problems in the operation of a lunar roving vehicle on the lunar surface, the astronauts remarked that the main concern would not be the consistency and mechanical behavior of the soil, but rather the general topography, and roughness characteristics of the lunar surface, the human tendency to underestimate distances, and the poor ability to discern the local vertical and to have a good definition of the horizon. All these factors would tend to restrict the operational speed of a lunar roving vehicle. However, no insurmountable problems appear to exist.

Soil Mechanics Observations at the LRL

The Soil Mechanics Team observed the properties of the lunar soil from the documented, core-tube, bulk, and contingency samples in the vacuum chamber and in nitrogen cabinets. In addition, the team performed a specific-gravity test and determined the density and resistance to penetration of fine-grained lunar material placed loosely in a container and also compacted in layers.

Vacuum Chamber

The outsides of the sample boxes were cleaned, and the boxes were then introduced into a special vacuum chamber, which is equipped with a set of spacesuit-type arm-length gloves that permit a technician to process and handle the samples under vacuum conditions. The gloves consist of an inner set and an outer set separated by low-pressure nitrogen gas (at 1 to 2 μm) that continually leaks into the chamber, limiting the

FIGURE 4-30. — Lunar sample in the vacuum chamber. (LRL photograph S-69-45016)

chamber pressure to approximately 10^{-8} T. This pressure is significantly greater than the pressures used in high-vacuum work in the study of ultraclean soil particles (refs. 4-16 to 4-19). However, water and organic molecules have been carefully eliminated from the system in order to produce a very clean vacuum.

The first sample box to be opened was the documented-sample box, which was the second one packed and sealed on the lunar surface. This sample box contained the roll of aluminum foil from the SWC experiment, the two core tubes, and approximately 20 lunar rocks. Except for the contents of the core tubes, the only soil in the documented-sample box was a small amount that had adhered to the rocks. Figure 4-30 shows one of these rocks placed on a viewing stand in the vacuum chamber. The positioning operation has shaken loose some of the soil. Microscopic examination of this soil revealed that it consisted of two basic sizes. The larger size fraction, comprising only 1 to 2 percent of the sample, consisted of clear, translucent, angular to subangular smooth particles that were 50 to 200 μm in size. The smaller size fraction consisted of brown, opaque particles, 2 to 10 μm in size, which occurred individually and in clusters. Because this soil had adhered to the rock, it is not necessarily representative of the lunar soil.

Core-Tube Samples

The core-tube samples were especially important because of the likelihood of being less disturbed than the bulk samples, which had been scooped up with the shovel, and also of retaining some of the stratigraphic structure, if any, of the lunar surface. The core tubes were transferred from the vacuum chamber to the Biological

Preparation Laboratory (BPL), where they were placed in one of the stainless-steel cabinets and surrounded by dry, sterile nitrogen gas at slightly less than atmospheric pressure. Core tube 2 contained a sample (fig. 4–31) 2.0 cm in diameter, 13.5 cm in length, and weighing 65.1 g, giving an average bulk density of 1.54±0.03 g/cc. The sample consists of a fine-grained granular material which was remarkably uniform in color, charcoal-gray with a slight brown tinge. There were fine reflecting surfaces over approximately 10 percent of the surface area, which gave the sample a sparkly appearance.

The phenomenon of the color difference between the surface and subsurface material observed by the astronauts and during Surveyor experiments was not confirmed in the core-tube samples. The sample had some cohesion and, therefore, retained its cylindrical shape after the top half of the split tube had been removed (fig. 4–32). In probing the sample, the action of the spatula indicated that the soil cohesion was slight, but was sufficient to hold small clumps of fines together. There appeared to be no variation in structure with length along the sample. A very slight color difference between the upper half and the lower half of the sample suggested some stratification, but probing revealed no discernible differences in the mechanical properties of the two halves.

About half of the sample, a total of 26.73 g, was removed along the length of the core (fig. 4–32) and placed in a nest of sieves. At this point it was found that the soil sample did not consist of fine particles only but contained a range of particle sizes. The larger particles were angular, dark, and glassy; some of the larger particles had vesicles in them. The very fine particles tended to form, break, and re-form lumps when shaken, as though the soil were slightly damp (fig. 4–33); although the grain-size distribution curve (fig. 4–34) indicates a fine, silty sand with a gravel trace, the aggregation of very fine particles during sieving may have biased the results toward the coarser size range.

Core tube 1 (fig. 4–35) contained a smaller sample, which was 2.0 cm in diameter, 10.0 cm in length, weighing 52.0 g, and giving an average bulk density of 1.66 ± 0.03 g/cc (fig. 4–36). The surface color of the soil was essentially the same as that of the soil in core tube 2. However, this sample contained numerous small cracks and voids that were not taken into account when computing its average bulk density. The soil was found to have the same consistency as the sample in core tube 2. Its grain-size distribution was determined by sieving, and the results have been plotted in figure 4–34.

For both core samplers, the core bit was flared inward at 15°, i.e., in the opposite direction from most terrestrial samplers. Thus, the soil was probably deformed considerably during sampling and the measured bulk densities are not necessarily indicative of the inplace lunar soil bulk density.

A specific-gravity test was performed on the material that had been removed from the core tubes for the sieve analyses. By using a gas comparison pycnometer, the total volume of the individual grains was determined to be 15.75 cc; and the total weight was 49.12 g. These values yield a specific gravity of 3.1, which is considerably higher than the typical value of 2.7 for terrestrial soils. The difference is partially attributed to the large amount of titanium oxide (6 percent) found in the core-tube soil sample.

Having determined the specific gravity of the solid particles, the void ratio e of the soil in the core tubes can be calculated from the following equation:

$$e = \frac{G\gamma_w}{\gamma} - 1 \qquad (4\text{-}2)$$

where G is the specific gravity of solids, γ_w is the density of water, and γ is the bulk density of the soil.

Substituting in equation (4-2) the measured values of G and γ for the soil samples in core tube 1 and core tube 2, and letting $\gamma_w = 1$ g/cc, the void ratio e_1 for the soil in core tube 1 is 0.87, and the void ratio e_2 for the soil in core tube 2 is 1.01.

Similarly, the porosity n can be determined from the following expression:

$$n = \frac{e}{1+e} \qquad (4\text{-}3)$$

By using equation (4-3) the porosity n_1 of the soil in core tube 1 is 46.5 percent and the porosity n_2 of the soil in core tube 2 is 50.1 percent.

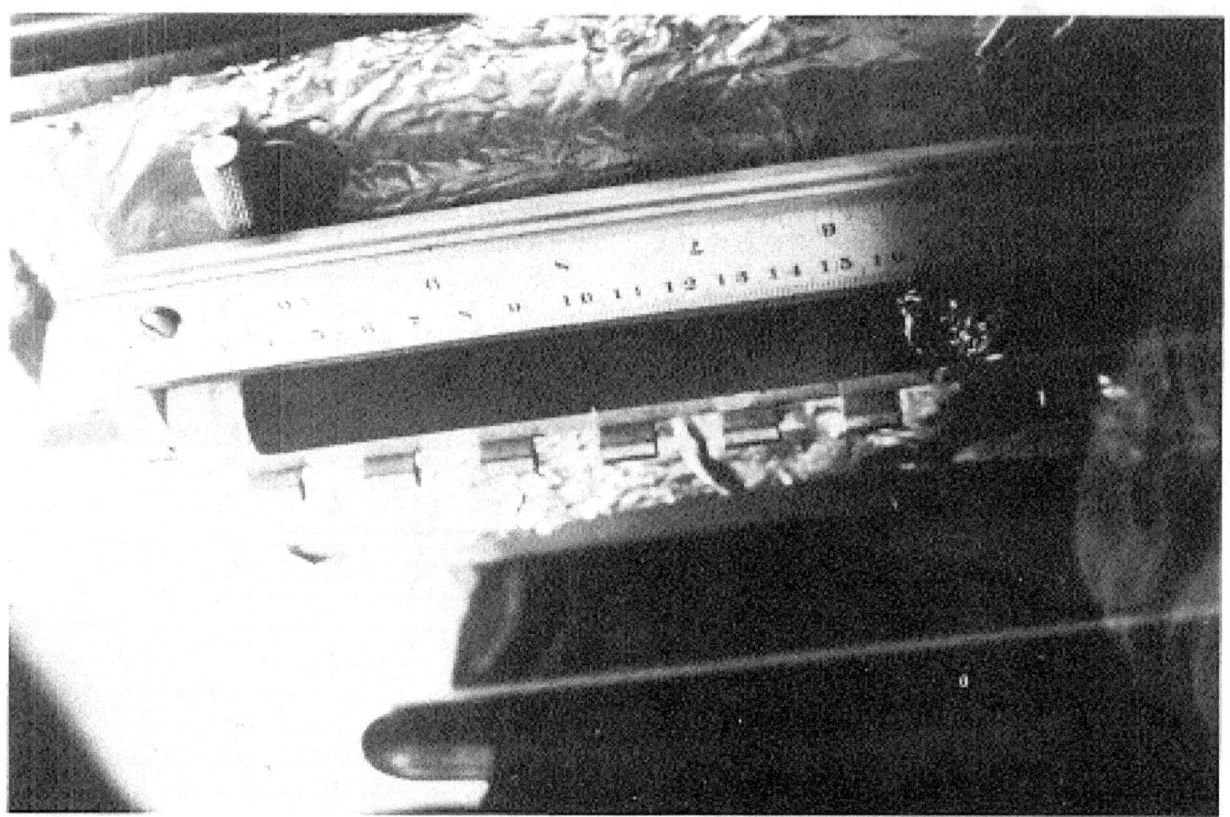

FIGURE 4-31. — First core tube (core tube 2) opened in the Biological Preparation Laboratory. (LRL photograph S-69-45930)

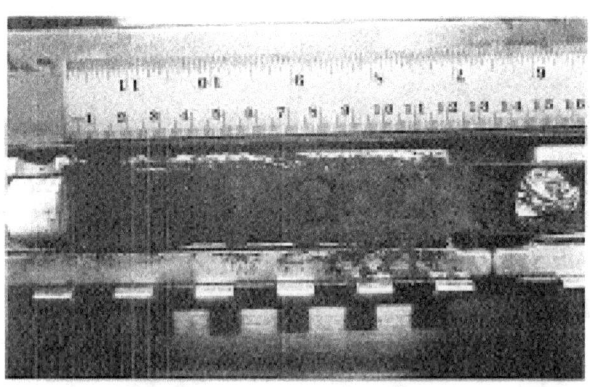

FIGURE 4-33. — Soil retained on sieve 140 (0.105 mm) from core-tube-2 sample. (LRL photograph S-69-45107)

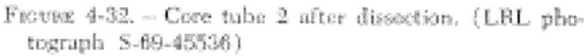

FIGURE 4-32. — Core tube 2 after dissection. (LRL photograph S-69-45536)

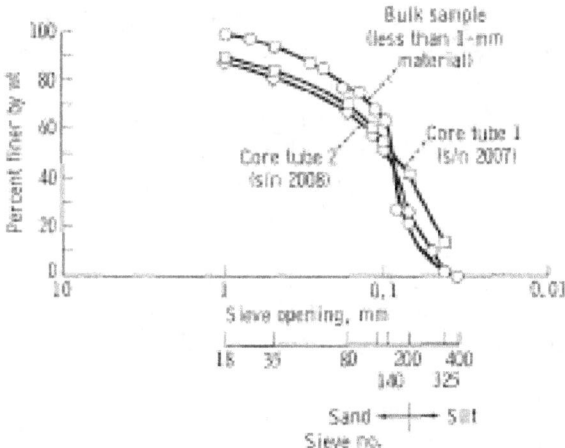

FIGURE 4-34. — Grain-size distribution curves of returned lunar soil samples.

FIGURE 4-35. — Lunar soil sample contained in core tube 1. (LRL photograph S-69-45048.)

Contingency Sample

The contingency sample was the first surface sample that was collected on the Moon within minutes after he stepped off the LM footpad. Armstrong scooped up the lunar material in fishnet style, discarded the contingency-sampler handle, folded over the open end of the Teflon bag, and put the sample bag into a pocket that was sewn to the outside of his spacesuit. This sample was given the name contingency sample because if the mission had been terminated, at least 500 g of lunar material would have been returned to Earth.

The contingency-sample bag was not vacuum sealed, as were the bulk-sample boxes and the documented-sample boxes, and the bag was exposed to the atmosphere in the LM. However, the behavior of the soil in the contingency sample was essentially the same as the behavior of other soil samples.

Although no specific-gravity determinations were made on the fine-grained material, several of the rocks were analyzed by the Preliminary Examination Team by means of a gas comparison pycnometer. Specific-gravity values ranging from 3.2 to 3.4 were obtained; these were slightly higher than the value obtained by the Soil Mechanics Team for the specific gravity of the soil from the core tubes. However, it should be noted that the volume of the rocks tested varied from

FIGURE 4-36. — Soil particles from the gas reaction sample. (LRL photograph S-69-45181.)

2.5 to 5 cc. Previous experience with the type of pycnometer used has shown that at least 10 cc are necessary for accurate results; thus, the specific gravity value for the rocks should not be considered definitive.

Gas Reaction Sample

The gas reaction sample consisted of fines taken from the bottom of the documented-sample box. This sample was examined microscopically and revealed the presence of small spherical particles mentioned previously. Although some metallic spheres were found, these particles consisted primarily of glasses. Selected glass spheres and mineral grains are shown in figure 4-36; the background in the photograph is an aluminum pan. The lunar soil grain shape can be seen to vary within a wide range from angular to well rounded.

Bulk Sample

The bulk sample was collected by Astronaut Armstrong during EVA after the contingency sample and prior to collection of the documented sample. The bulk sample consisted of rock and soil samples that were scooped up from the lunar surface and were sealed in a separate vacuum box. The bulk-sample box remained sealed in another part of the vacuum chamber in the LRL while the documented-sample box was opened and its contents were examined.

While the documented samples were being processed, a leak developed in one of the vacuum gloves; as a consequence, the bulk-sample box, which had retained its integrity, was moved to the cabinets in the Biological Preparation Laboratory. There the bulk sample was processed in dry, sterile nitrogen at atmospheric pressure.

The relative ease of working in these cabinets, and the capability to sustain simultaneous operations by several operators, provided an opportunity to perform simple tests that would have been impossible to perform in the vacuum chamber. These tests were the determination of the bulk densities of loosely deposited and compacted lunar soil as well as the resistance to penetration of the soil at these densities. Although more extensive testing would have been preferred, under the circumstances the data are extremely valuable.

Before performing these tests, the Soil Mechanics Team observed the processing of the bulk sample, beginning with the separation of the 30 lb of rock and soil into three sizes of fractions (coarser than 10 mm, coarser than 1 mm, and finer than 1-mm size). As the sample was being transferred from the box to the set of sieves, using the scoop, clean, smooth vertical walls approximately 3 in. deep and similar in appearance to the trenches dug in the lunar surface by Surveyor 3 and Surveyor 7 were cut in the soil mass. This characteristic behavior indicates the significant amount of cohesion of the lunar soil, which apparently is unaffected by exposure to nitrogen gas at atmospheric pressure. However, it should be noted that the atmosphere in the cabinets is not pure nitrogen, but contains oxygen, water vapor, and organic molecules. On the other hand, because the minerals in the soils and on the rocks were found by geologists to be remarkably unweathered, it is, therefore, to be expected that long-term exposure of the lunar materials to the atmosphere in the nitrogen cabinets will alter their properties.

The cohesion and the dark charcoal-gray brownish color of the lunar soil masked the presence of even sizable rocks within the soil mass. When a scoopful of soil that appeared to consist of only fine-grained material was placed on the 10-mm sieve, several fragments approximately 3 cm in size were retained as the soil fell through the sieve mesh. This masking property may explain why only few rocks appeared to have been excavated by Surveyor 3 and Surveyor 7. On the other hand, the lunar soil does not appear to be significantly adhesive; in most cases, it formed only a fine dust layer on stainless-steel tools, although in some instances removal of lunar soil that had adhered to the sides of the sieve could only be accomplished by scraping off the soil.

Penetration tests were performed on the soil fraction with grains finer than 1 mm in size. The grain-size distribution of this soil is shown in figure 4-34. The soil was first placed as loosely as possible into a sample can (with a height of 5.1 cm and a diameter of 8.85 cm), and its bulk density was determined to be 1.36 ± 0.01 g/cc. This density corresponds to a void ratio of 1.28 and a porosity of 56 percent (table 4-II). The minimum density of the lunar soil at $\frac{1}{6}$ g should be less than this value, because a looser soil structure would develop as a result of the adhesive forces acting between the soil particles.

TABLE 4-II. *Packing characteristics of loosely deposited and compacted lunar soil*

State	Density, g/cc	Void ratio[a]	Porosity, percent[a]
Loose	1.36 ± 0.01	1.28	56.0
Compacted	1.80 ± 0.01	.72	41.8

[a] Calculated from measured specific gravity of 3.1.

A simple penetrometer was pushed into the surface of the loose soil sample five times. In four tests, the ⅛-in.-diameter penetrometer did not meet with sufficient resistance to compress the spring. During the fifth test, the back end of the penetrometer (with an area of 0.416 in.^{2}) was used and the force and penetration were measured. Pertinent data are shown in table 4–III and in figure 4–37. In all five tests the soil failed in a punching mode, as can be seen in figure 4–38.

In a second series of tests the soil was placed in the sample can in several layers and was compacted by rodding, tapping, and compressing. In this dense state the bulk density of the soil was determined to be 1.77±0.01 g/cc, corresponding to a void ratio of 0.75 and a porosity of 42.8 percent. Six penetration tests were conducted, and the data are shown in table 4–III and figure 4–37. The data show a considerable increase in the penetration resistance of the compacted lunar soil. In addition, the failure mode was one of classical, incompressible shear with well-defined radial and circumferential cracks and heaving of the surrounding soil (fig. 4–39).

The soil was compacted again in an attempt to achieve a denser state. The resulting bulk density of the soil was determined to be 1.80±0.01 g/cc, corresponding to a void ratio of 0.72 and a porosity of 41.8 percent. A two-stage penetration test was performed which indicated a further increase in the penetration resistance due to compaction. Figure 4–39 shows the results of this test.

Preliminary analysis of the penetration data has revealed the following:

(1) The resistance to penetration of the dense soil is more than 20 times greater than the resistance to penetration of the loose soil.

FIGURE 4-38. — Penetration of loose lunar soil sample. (LRL photograph S-69-47484)

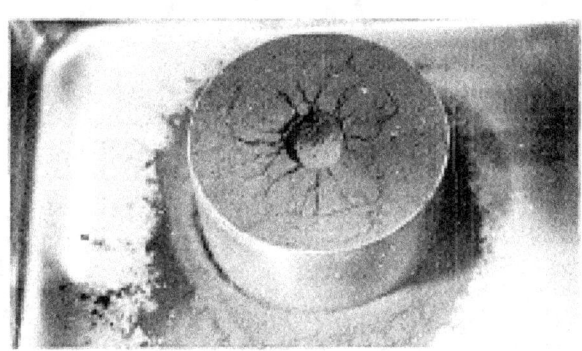

FIGURE 4-39. — Penetration of compacted lunar soil sample. (LRL photograph S-69-47489)

(2) The cohesion for the dense soil ranges from 0.05 to 0.20 psi.

(3) The cohesion does not appear to be affected by short-term exposure to nitrogen and trace amounts of oxygen, water, and organic materials at atmospheric pressure.

(4) If the lunar surface soil in the LRL nitrogen cabinet is compacted to a density close to the maximum bulk density of the returned sample, the soil offers sufficient resistance to penetration to account for the behavior observed by the astronauts when they pushed various tools into the lunar surface material.

Summary and Conclusions

The upper centimeters of the surface material in the vicinity of Tranquility Base are characterized by a brownish, medium-gray, slightly cohesive granular soil, which is largely composed of bulky grains in the silt to fine-sand size range.

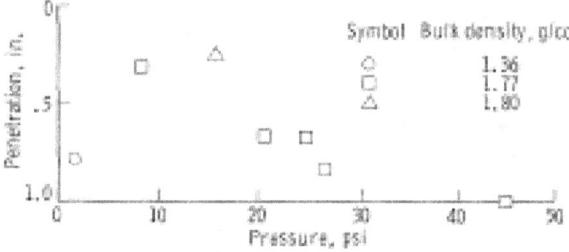

FIGURE 4-37. — Penetration versus resistance to penetration of loose and compacted lunar soil sample.

TABLE 4-III. *Lunar soil bulk sample. Penetration test results*

Test number	Density, g/cc	Force, lb	Area, in.²	Pressure, psi	Penetration, in.	Pressure-penetration ratio, psi/in.
1	1.36	<0.41	0.049	<8.3	0.25	—
2	1.36	<.41	.049	<8.3	.77	—
3	1.36	<.41	.049	<8.3	.77	—
4	1.36	<.41	.049	<8.3	.77	—
5	1.36	.69	.416	1.65	.79	2.1
6	1.77	<.41	.049	<8.29	.32	26.3
7	1.77	1.22	.049	24.8	.67	37.0
8	1.77	.41	.049	8.29	.25	—
9	1.77	2.20	.049	44.7	1.00	44.7
10	1.77	1.30	.049	26.5	.83	31.9
11	1.77	8.70	.416	20.7	.67	30.9
12(a)	1.80	6.50	.416	15.6	ᵃ.26	60.9
12(b)	1.80	17.95	.416	43.1	ᵃ.77	55.9

ᵃ The penetrometer was pushed into the soil to a maximum penetration of 0.26 in.; then it was removed and applied again at the same place until a total penetration of 0.77 in. was recorded from the original surface of the sample.

Angular to subrounded rock fragments are distributed throughout the area. Some of these fragments are lying on the surface, some are partly buried, and some are barely exposed.

The lunar surface is relatively soft to depths ranging from 5 to 20 cm. The soil can be easily scooped, offers low resistance to penetration, and provides low lateral support for staffs, poles, or core tubes. Beneath this relatively soft surface, the resistance of the material to penetration increases considerably.

The mechanical behavior of the lunar soil can be summarized as follows:

Confinement of the loose material leads to a significant increase in resistance to deformation, which is a characteristic of soils deriving a large portion of their strength from interparticle friction.

The soil has a small amount of cohesion. This was evidenced by the following observations:

(1) The soil has the ability to stand on vertical slopes and to retain the detail of a deformed shape. The sidewalls of trenches dug with the scoop were smooth with sharp edges.

(2) The fine grains stick together and, in some locations, the astronauts found it difficult to distinguish soil clumps from rock fragments.

(3) The holes made by the core tubes were left intact upon removal of the tubes.

(4) The material collected at the bottom of the core tubes did not pour out when the core bit was unscrewed.

Natural clods of fine-grained material crumbled under the astronauts' boots. This may be indicative of some cementation between the grains, although LRL tests indicated that the soil grains tended to cohere again after being separated.

Most of the footprints indicate that the low unit loads of approximately 0.7 N/cm² (1 psi) imposed by the astronauts caused compression of the lunar surface soil, although bulging and cracking of the soil adjacent to a footprint occurred in a few instances. The latter observation indicates shearing rather than compressional deformation of the soil.

The lunar surface soil was eroded by the LM DPS engine; some of the particles removed during this erosion were observed to travel a considerable distance.

Available information indicates that the LM landing was achieved under essentially static loading conditions as far as the landing gear was concerned. The relatively small LM footpad penetrations of 2.5 to 7.5 cm (1 to 3 in.) correspond to average static-bearing pressures of 0.6 to 1.5 N/cm² (0.8 to 2.1 psi).

Granular material from the surface adhered to various objects such as the astronauts' boots and clothing, the TV cable, and the LEC. It was possible, however, to jar the sand-sized particles loose from the boots.

Astronaut mobility was not significantly hampered by the soil conditions encountered; however, irregular topography may pose control problems in the operation of a roving vehicle during subsequent lunar missions.

Preliminary visual and microscopic examination at the LRL of the silt-sized particulate material adhering to and detached from a lunar rock in the vacuum chamber showed that:

(1) The majority of the individual particles were dark, opaque, blocky, angular, and fresh in appearance.

(2) A small proportion of the grains, generally those grains larger in size (up to fine-sand size), were clear, translucent, blocky, and angular.

(3) Groups or clumps of particles were present.

(4) Shards, needles, or filaments were not observed among the particle shapes.

Visual examination of the material in the core tubes at the LRL indicated the following:

(1) The sample was remarkably uniform in its charcoal-brown color.

(2) The soil had the appearance of being homogeneous and uniform in grain size.

(3) Small reflecting surfaces gave the sample a sparkly appearance.

(4) Cohesion was evident in that both core-tube samples retained their cylindrical shape.

(5) One sample was much more disturbed in appearance than the other, and the sample with the disturbed appearance exhibited fissures and cracks.

Sieve analyses were performed at the LRL on the fine-grained material collected with the documented sample, the core tubes, and the bulk sample. In all three situations, the grain-size distribution was found to be that of a silty-fine sand. However, the aggregation of individual particles biased the analysis toward the larger size range.

At the LRL the specific gravity of lunar soil was measured to be 3.1, which is considerably higher than the average value of approximately 2.7 for terrestrial soils. Based on the specific gravity obtained for the lunar soil and on the measured bulk densities, the void ratio of the material in core tube 1 is 0.87 and in core tube 2 is 1.01. The porosities in core tubes 1 and 2 are 46.5 and 50.1 percent, respectively. Because of the disturbance involved in sampling, these values may not be representative of the inplace properties of the material.

At the LRL material finer than the 1-mm size obtained from the lunar bulk sample was placed loosely in a container, and the bulk density of the material was found to be 1.36 g/cc. When compacted, the bulk density of the lunar soil was found to be 1.80 g/cc. In the compact state, the bearing capacity of the material was determined by a small penetrometer. From these tests, the cohesion of the material was estimated to be in the range between 0.04 and 0.14 N/cm² (0.05 and 0.20 psi). These experiments were performed in the nitrogen atmosphere of the cabinets in the Biological Preparation Laboratory.

References

4-1. Kopal, Z.: Physics and Astronomy of the Moon. Academic Press, 1962.

4-2. Jaffe, L. D.: Lunar Dust Depth in Mare Cognitum. J. Geophys. Res., vol. 71, no. 4, Feb. 1966, pp. 1095-1103.

4-3. Jaffe, L. D.; and Scott, R. F.: Lunar Surface Strength: Implications of Luna 9 Landing. Science, vol. 153, July 22, 1966, pp. 407-408.

4-4. Christensen, Elmer M.; Batterson, S. A.; et al.: Lunar Surface Mechanical Properties — Surveyor I. J. Geophys. Res., vol. 72, no. 2, June 15, 1967, pp. 801-813.

4-5. Cherkasov, I. I.; et al.: Determination of the Physical and Mechanical Properties of the Lunar Surface Layer by Means of Luna 13 Automatic Station. Moon and Planets II. North-Holland Publishing Company, 1968, pp. 70-76.

4-6. Scott, Ronald F.: The Density of the Lunar Surface Soil. J. Geophys. Res., vol. 73, no. 16, Aug. 15, 1968, pp. 5469-5471.

4-7. Quaide, William L.; and Oberbeck, Verne R.: Thickness Determinations of the Lunar Surface Layer from Lunar Impact Craters. J. Geophys. Res., vol. 73, no. 16, Aug. 15, 1968, pp. 5247-5270.

4-8. Scott, R. F.; and Roberson, F. I.: Soil Mechanics Surface Sampler: Lunar Surface Tests and Analyses. Chap. IV of Surveyor III Mission

4-9. CHRISTENSEN, E. M.; et al.: Lunar Surface Mechanical Properties. Chap. IV of Surveyor V Mission Report, Part II: Science Results, TR32-1246, Jet Propulsion Laboratory, Calif. Inst. Tech., Nov. 1, 1967, pp. 43-89.

4-10. CHRISTENSEN, E. M.; et al.: Lunar Surface Mechanical Properties. Surveyor VI Mission Report, Part II: Science Results, TR32-1262, Jet Propulsion Laboratory, Calif. Inst. Tech., Jan. 10, 1968, pp. 47-107.

4-11. SCOTT, R. F.; and ROBERSON, F. I.: Soil Mechanics Surface Sampler. Chap. V of Surveyor VII Mission Report, Part II: Science Results, TR32-1264, Jet Propulsion Laboratory, Calif. Inst. Tech., Mar. 15, 1968, pp. 135-185.

4-12. HUTTON, R. E.: LM Soil Erosion and Visibility Investigations, Part I: Summary Report and Part II: Appendix (References). TRW Tech. Rept. 11176-6060-RO-00, Aug. 13, 1969.

4-13. SCOTT, RONALD F.; and KO, HON YIM: Transient Rocket-Engine Gas Flow in Soil. AIAA, vol. 6, no. 2, Feb. 1968.

4-14. LAND, NORMAN S.; and CLARK, LEONARD V.: Experimental Investigation of Jet Impingement on Surfaces of Fine Particles in a Vacuum Environment. NASA TN D-2633, Apr. 1965.

4-15. MEYERHOF, G. G.: The Ultimate Bearing Capacity of Foundations. Geotechnique, vol. II, no. 2, Dec. 1951, pp. 301-330.

4-16. HALAJIAN, J. D.: The Case for a Cohesive Lunar Surface Model. Rept. no. ADR-04-04-64.2, Grumman Aircraft Engineering Corp., 1964.

4-17. BROMWELL, LESLIE G.: The Friction of Quartz in High Vacuum. Research Rept. R66-18, Department of Civil Engineering, Soil Mechanics Division, MIT, May 1966.

4-18. RYAN, J. A.: Adhesion of Silicates in Ultrahigh Vacuum. J. Geophys. Res., vol. 71, no. 18, Sept. 15, 1966, pp. 4413-4425.

4-19. NELSON, JOHN D.; and VEY, E.: Relative Cleanliness as a Measure of Lunar Soil Strength. J. Geophys. Res., vol. 73, no. 12, June 15, 1968, pp. 3747-3764.

5. Preliminary Examination of Lunar Samples

This report describes the results of preliminary physical, chemical, and biological analyses of the lunar material returned from the July 20, 1969, Apollo 11 lunar landing at Mare Tranquillitatis. The analyses were performed at the Lunar Receiving Laboratory (LRL) by the Preliminary Examination Team (PET) consisting of resident NASA staff members and visiting scientists. The results of these analyses were used to plan the allocation and distribution of the lunar materials to the 142 scientists throughout the world who will conduct detailed studies of the samples.

The Apollo 11 astronauts collected three separate samples: a contingency sample, a bulk sample, and a documented sample. The contingency sample, weighing approximately 1 kg, was collected within 1.5 m of the lunar module (LM) and consisted of a few rock fragments and fine materials. The bulk-sample box contained approximately 14 kg of rocks and fine materials and was collected within 10 m of the LM. The documented-sample box contained approximately 6 kg of rocks selected late in the extravehicular activity (EVA) and two core tubes.

Both the bulk-sample and the documented-sample containers were sealed on the lunar surface, transferred into the LM, bagged, and (after rendezvous) transferred into the command and service module (CSM). The sample containers were removed from the command module (CM) on the carrier Hornet, transferred into the Mobile Quarantine Facility (MQF) and placed in a second bag, and (after decontamination) transferred from the MQF into two aircraft that returned the containers to the LRL on July 25, 1969. The lunar samples in the sample boxes were brought into the Sample Laboratory (SL) of the LRL and put into decontamination chambers, where terrestrial contamination was removed by a peracetic acid spray.

The biological barrier in the SL is actually a double barrier. The primary barrier consists of vacuum chambers and special class III biological cabinets, inside which the samples are handled. The sealed walls of the building constitute a secondary barrier. The inside of the Crew Reception Area (CRA) and the SL are operated at a small negative pressure so that any leakage will be inward. Similarly, SL primary barrier cabinets are operated at negative pressure with respect to the rooms in which they are located. The biological barriers are two-way barriers that prevent unsterilized lunar material from contaminating the atmosphere of the Earth and prevent terrestrial contaminants from reaching the samples.

All samples were opened and processed in a vacuum of 5×10^{-6} T (or less) or in a dry nitrogen atmosphere. All handling was done in gloved cabinets. The documented-sample box was opened in the vacuum chamber, and the rocks were examined, described, photographed, weighed, and chipped for physical and chemical analysis and biological testing. The bulk-sample box, the contingency-sample container, and the core tubes were opened in a dry nitrogen atmosphere inside one of the special cabinets after proper determination that no short-term degradation of the sample occurred upon exposure to oxygen, moist air, and carbon dioxide. Approximately 50 g of the core-tube material and 500 g of the fines from the bulk box were removed for biological testing. Most of the remaining material was stored in a nitrogen atmosphere inside the biological cabinets; the rest was stored under vacuum conditions.

Portions of the sample were used for determination of density, of reaction to gas, of mineralogical properties, chemical analysis by optical emission spectroscopy, total organic carbon analysis, and X-ray diffraction properties in the Physical-Chemical Test Laboratory of the LRL. Rare gases and volatile organic species were analyzed in the Gas Analysis Laboratory. Counting of short-lived isotopes was performed in the LRL Radiation Counting Laboratory, a specially shielded underground laboratory. Magnetic properties and properties of the soil were also measured. The results of these studies are the subject of this report.

Mineralogy and Petrology

The total weight of the lunar material returned by Apollo 11 was 22 kg, of which 11 kg were rock fragments more than 1 cm in diameter and 11 kg were smaller particulate material. Because the documented sample container was filled by picking up selected rocks with tongs, the container held a variety of large rocks (total weight 6.0 kg). The total weight of the bulk sample was 14.6 kg.

The following discussion is based on stereomicroscopic examination of the samples, aided by random grain counts under the polarizing microscope. Some of the provisional identifications were reinforced by limited further investigation with the aid of spindle-stage and X-ray powder diffraction methods, together with study of thin sections of two of the crystalline rocks which became available near the close of the preliminary examination. In spite of the handicap of an adherent layer of dust, all the rocks were examined. Only the contingency samples and small chips taken from the samples in the rock boxes were free of dust when examined.

The returned lunar material may be divided into the following four groups:

(1) Type A — fine-grained vesicular crystalline igneous rock
(2) Type B — medium-grained vuggy crystalline igneous rock
(3) Type C — breccia consisting of small fragments of gray rocks and fine material
(4) Type D — fines

The term "rocks" is applied to fragments larger than 1 cm in diameter; the term "fines" is applied to fragments smaller than 1 cm in diameter. All the rocks and many smaller rock fragments show unearthly surface features (fig. 5-1) that are discussed in the following paragraphs.

The crystalline rocks are volcanic in origin.

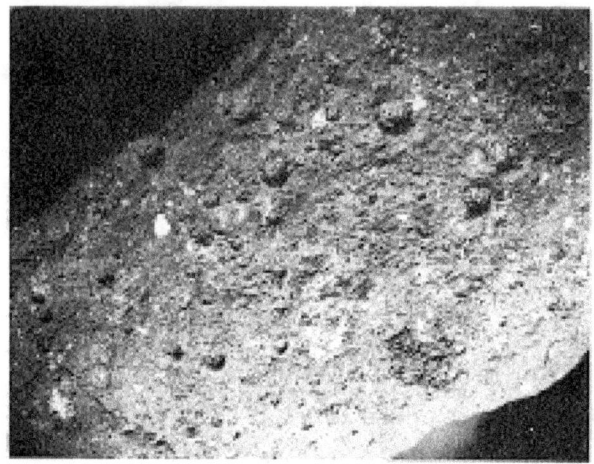

FIGURE 5-1. — Surface pits in breccia.

The term "volcanic," as used in this report implies surface lavas or near-surface igneous rocks and carries no connotations regarding impact-generated or impact-triggered volcanism versus volcanism in the common terrestrial sense.

The rocks contain pyrogenic mineral assemblages and gas cavities indicating crystallization from melts. The major minerals can be assigned to known rock-forming mineral groups. The unique chemistry of the magmas has resulted in mineral ratios unlike those of known terrestrial volcanic liquids, yet not greatly different from some terrestrial cumulates, at least in the major elements.

Twenty crystalline rocks, most of which weigh more than 50 g, were returned. The largest crystalline rock weighed 919 g. These rocks have been divided into a fine-grained vesicular type (type A, fig. 5-2) and a coarser grained vuggy microgabbroic type (type B, fig. 5-3), but the rocks may be members of a textural and compositional series. Figures 5-2 and 5-3 illustrate the two most common types.

A chip from a dark-gray vesicular rock with subophitic texture (type A) has a bulk density of approximately 3.4 g/cc. The vesicles range between 1 and 3 mm in diameter, and most of the vesicles are spherical; however, some are ovate. Coalescence has modified other vesicles to create irregularly shaped cavities that are larger than single vesicles. The vesicles are faced with brilliant reflecting crystals composed of the ground-

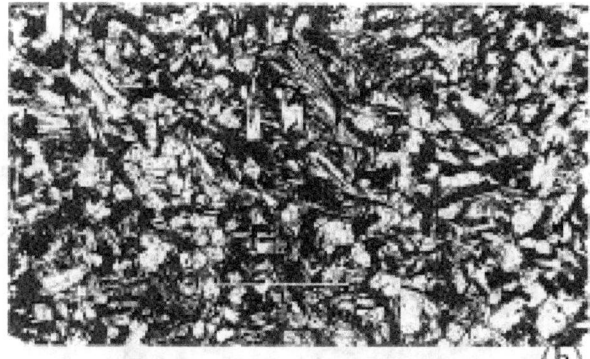

FIGURE 5-2.—Typical type A crystalline rock. (a) Sample 10022. (b) Photomicrograph of sample 10022.

mass minerals. Such vesicles contain no transecting crystals or sublimate minerals. In addition to the spherical vesicles, minor irregularly shaped cavities (or vugs) exist into which groundmass and accessory minerals project locally. Vugs tend to take the place of spherical vesicles in the more coarse-grained rocks; euhedral crystals projecting into vugs are common.

Preliminary modal analysis of this type A specimen, recalculated for 15 percent of void space, yields clinopyroxene, 53 percent; plagioclase, 27 percent; opaques (abundant ilmenite, perhaps minor troilite, and perhaps native iron), 18 percent; other translucent phases (at least two), 2 percent; and minor olivine. Notable are rare grains of olivine showing marginal transformation to clinopyroxene. The olivine grains are as much as 0.5 mm long, whereas most other mineral grains range between 0.05 and 0.2 mm in diameter. In vuggy regions, there is a slight increase in grain size. Except for the high con-

FIGURE 5-3.—Typical type B crystalline rocks. (a) Sample 10047. (b) Sample 10050.

tent of opaque minerals, the lunar rock resembles some terrestrial olivine-bearing basalts. Mineral grain and vesicle sizes, shapes, and distribution suggest that the rocks originated near the top or bottom of a lava flow or lava lake.

Although most of the type A rocks have smaller vesicles, nine other type A rocks are similar to the rocks previously described. The most notable variation among type A rocks is in olivine content, which ranges from zero in some rocks to approximately 10 percent in other rocks.

The dark-brownish-gray, speckled type B

rock (fig. 5-3) has irregular cavities and a bulk density of approximately 3.2 g/cc. The texture is granular and in general appearance resembles the microgabbroic textures of segregation veins and pods in some terrestrial basalts.

The grain size in type B rocks varies from 0.2 to 3.0 mm. The largest crystals, many of which are euhedral, project into cavities. Preliminary modal analysis indicates clinopyroxene, 46 percent; plagioclase, 31 percent; opaques (mainly ilmenite), 11 percent; low cristobalite, 5 percent; and others, 7 percent. The others include an unidentified yellow mineral that seems to be concentrated in vuggy areas of the rock and a high-index colorless mineral. Olivine is not present in this rock or in similar rocks.

The crystalline rocks present a series based on groundmass grain size ranging from approximately 0.1 to 1 mm, with larger crystals (up to 3.0 mm) in vuggy zones of the coarser grained rocks. Ilmenite seems less abundant, and low cristobalite seems more abundant in the coarser grained rocks. In general, the abundance of vugs increases and the abundance of vesicles decreases as grain size increases. Olivine is found only in the finer grained rocks, and the unidentified yellow phase is found only in the coarser grained rocks. The three major minerals (clinopyroxene, calcic plagioclase, and ilmenite) are present in all the rocks.

The apparently complete absence of hydrous mineral phases is notable, as is the extremely fresh appearance of all the crystalline rock interiors, in spite of the microfractures and high potassium-argon, which suggests that the rocks are old.

All the breccias examined (type C, fig. 5-4) are mixtures of fragments of different rock types and are gray to dark gray, with specks of white, light-gray, and brownish-gray rock fragments. Most breccias are fine grained, with fragments smaller than 1 cm in diameter; fragments smaller than 0.5 cm predominate. Only a few fragments in the breccias are rounded. The fragments consist of either rocks or minerals similar to those described previously, but have the distinction that many fragments show a greater population of closely spaced microfractures and various degrees of vitrification. In addition, angular fragments and spherules of glass (which also

(a)

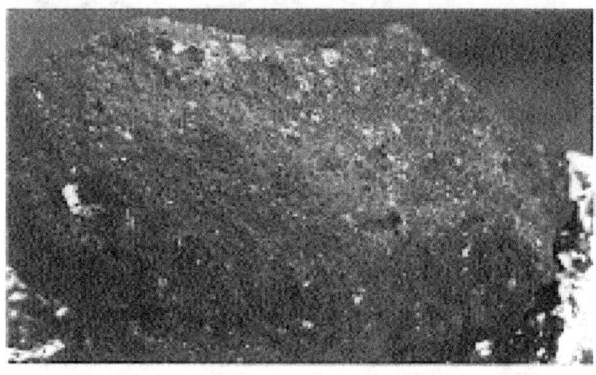

(b)

(c)

FIGURE 5-4. — Type C breccia. (a) Sample 10019 illustrating rounding and glass-lined pits. (b) A chip of sample 10019 illustrating rounding, glass-lined pits, and lithic fragments. (c) A chip of sample 10019 illustrating rounding and lithic fragments.

are characteristic components of the lunar fines) are present in a wide range of color and refractive indices. In several specimens, vesicular fine-grained crystalline rock constitutes a major fraction of the larger fragments. In other specimens, mineral fragments and small fragments of the coarser microgabbroic type predominate. Although the rock fragments are small, there seems to be a wider range of variation in modal mineral composition and in shock metamorphism of the crystalline rock types than is shown by the large crystalline rock samples returned. Several minor rock types, one predominantly light gray and one predominantly pale yellow, occur in breccia but are not represented among the larger individual rocks. Some breccias contain crystal fragments larger than 3 mm across, indicating the existence of coarser crystalline rocks than those returned.

The finer fraction of the breccias is gray and granular in appearance and consists largely of glass particles (both angular and rounded) and fragments of birefringent minerals, which are coated in part with glass or with opaque or fine-grained dusty material. The glass particles, which form most of the matrix, are similar to the glass of the finer material. Many single glass particles are composed of more than one glass type and are therefore unlike glass shards of common terrestrial volcanic processes. Some of the breccias are transected by vesicular glass veins or contain pods of glass, either formed in place or injected along fractures.

The degrees of induration and the histories of subsequent deformation of the breccias are varied. While some breccias are poorly consolidated and are soft and friable, others have coarse layering. Still other breccias have closely spaced fracture systems and are as hard as (if not harder than) the hardest of the crystalline rocks. The breccias and many of the broken fragments of crystalline rocks in the breccias are composed of impact ejecta.

Two types of unique surface features occur on all rocks of the lunar samples: small pits lined with glass, and glass spatters not necessarily associated with pits. In addition, the crystalline rocks show a generally lighter colored surface, compared to the interior, which appears to be related to microfracturing of surface crystals.

The diameters of pits average somewhat less than 1 mm. Diameter-to-depth ratios of the pits have a range of values, but the ratio is apparently smaller for pits in the breccias than for pits in the crystalline rocks. A few of the rocks examined show pitting on rounded sides but no evident pitting on one (generally flat or irregularly rough) surface. The glass surfaces in the pits are bright reflecting and are commonly uneven and botryoidal. Botryoidal surfaces are more common in pits within breccias than in pits within crystalline rocks. Raised, glassy rims occur in greater abundance in surrounding pits within the breccias. The glass extends beyond some pit rims. Fractures and rare glass veinlets radiate from some pits. The pits are presumably caused by the impact of small particles on the surfaces of the rocks.

In addition to glassy pits, thin glass crusts occur that appear to be the result of spattering. Spatter crusts more than 1 cm in diameter occur on both breccia and crystalline rock surfaces. These crusts may be related to nearby impact events.

The surfaces of the crystalline rocks show whitish blotches and halos around the glassy pits. This whitish color is at least partially attributable to intense microfracturing of minerals, particularly feldspar and pyroxene, and does not penetrate more than 0.5 to 1.0 mm below the surface in most of the cross sections examined. In some crystalline rocks, surface whitening is so widespread that the whole surface is lighter in color than the interior. This feature is particularly noticeable in fine-grained rocks, which are dark gray on freshly broken surfaces.

The most noticeable surface feature of the rocks is the rounding of one or more edges and corners. The most striking example of rounding, and perhaps the most common, is that in which one side of a rock is nearly flat and the remainder of the surface is rounded. Rounding appears to be more pronounced in the softer, more friable breccias than in the crystalline rocks. The breccias commonly have coarser grains projecting above the original surface, which suggests that the finer grains that once surrounded them have been eroded away. Thus, the appearance is similar to that of friable sedimentary rocks of a wide range in grain size that have been sandblasted. Both the rounding and the detailed surface appearance indicate that an erosional process has acted on the rocks. The surface features of several rocks allow conclu-

FIGURE 5-5. — Sample 10046 illustrating a type B rock that displays evidence of its original position. The upper right portion of the sample displays rounding, glass-lined pits, and a light color. The lower left portion of the sample displays a sharp angular outline, few pits, and a dark color. This sample appears to have been partially buried on the lunar surface with only the upper right portion exposed.

sions to be made regarding the orientation of the rock on the lunar surface (fig. 5-5).

Two core samples of material 2 cm in diameter were returned: core tube 1 contained 10 cm and core tube 2 contained 13.5 cm of material. The core samples were composed predominantly of particles in the range of 1 mm to 30 μm, with admixed angular rock fragments, crystal fragments, glass spherules, and aggregates of glass and lithic fragments in the coarser sized fraction. When the upper half of the split-tube core liner was removed, the cylindrical shape of the material was retained perfectly. Both the material in the tubes and the fines are generally medium dark gray with a tinge of brown. When prodded with a small spatula, the material disintegrates particle by particle or forms extremely fragile ephemeral units of subangular blocky shapes.

Neither core sample showed obvious grain-size stratification. The core sample taken from core tube 2 displayed a slightly lighter zone approximately 6 cm from the top surface. This zone is 2 to 5 mm thick with a sharp upper boundary and a gradational lower boundary. In grain size or texture, the lighter zone is not megascopically different from the dark material.

Sieve analyses of material from the two core tubes, a sample from the documented Apollo lunar surface return container (ALSRC), and a sample from the bulk ALSRC, are shown in figure 5-6. These distributions are replicable by simple dry sieving, but may be biased by an aggregation of fines. Core tube 1 has a bulk density of 1.66±0.03 g/cc, and core tube 2 has a density of 1.54±0.03 g/cc.

The fines consist chiefly of a variety of glasses, plagioclase, clinopyroxene, ilmenite, and olivine. Rare spherules and rounded fragments of nickel-iron up to 1 mm in size were observed. The glass, which constitutes approximately one-half the material, is of three types: (1) botryoidal, vesicular, and globular dark-gray fragments; (2) angular pale or colorless (or more rarely, brown, yellow, or orange) fragments with a refraction index ranging from approximately 1.5 to approximately 1.6; and (3) spheroidal (fig. 5-7), ellipsoidal, dumbbell-shaped, and tear-drop-shaped bodies, most of which are smaller than 0.2 mm and range in color from red to brown and from green to yellow. Refraction indices range from less than 1.6 to more than 1.8 and are generally higher for the more intensely colored glass. Material with a refraction index greater than 1.7 is less common than material with a lower index. The angular colorless and pale glass, by far the more abundant type, is in part turbid or weakly birefringent. Unlike nor-

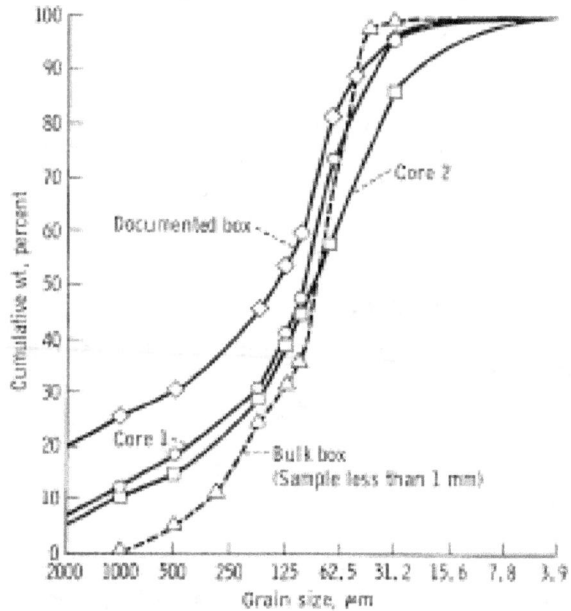

FIGURE 5-6. — Grain-size distribution in the lunar fines.

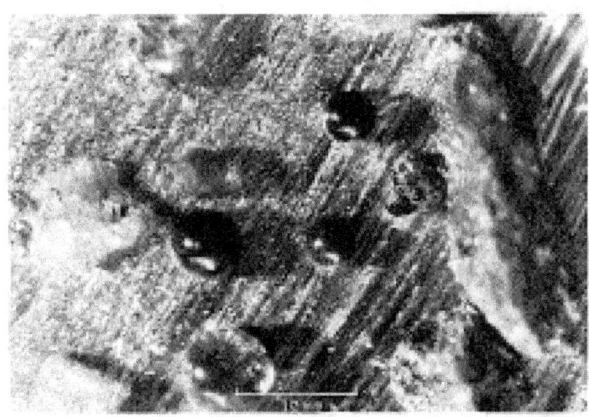

FIGURE 5-7.—Glass spherules of various colors found in fine-grained material.

mal quenched magma droplets and glasses from terrestrial volcanic sources, many single glass particles found in the core samples are inhomogeneous.

Evidence of impact metamorphism exists in the lunar samples, particularly in the loose, fine-grained materials and in the breccias. In contrast, most of the crystalline rocks, although commonly fractured or crumbly on the surface, show negligible or weak shock effects in their interiors. Only a small number show evidence of strong shock. Crystalline rocks underlying glass-lined pits are crushed or powdered, but are not strongly shocked. Such regions range from limited to wide distribution on the surfaces of the crystalline rocks.

Many phenomena were observed in the loose material and in the breccias that indicated that melting was induced by strong and intense shock. These phenomena included (1) glass dumbbells, teardrops, and other forms of evolution; (2) vesicular and flowed glass containing at least two types of glass; and (3) nickel-iron spherules. The abundance of vitrified mineral fragments is also evidence of moderate to strong shock. Unshocked minerals or lithic fragments are found in the loose materials and in breccia, but these fragments are not abundant. Most birefringent crystal fragments found in the fines display pronounced straight to mosaic undulatory extinction, and several display lamellar microstructures, indicating that these fragments have encountered weak to strong shock.

Evidence of a multiple shock history is provided by fragments of breccia within breccia and by breccias that contain spherules of glass from prior impact events and that are splashed with glass from subsequent impact events. Each breccia sample contains a wide variety of mineral and lithic fragments of various degrees of shock, and each sample appears to have a complex history.

Clinopyroxene is present in all of the rock samples examined. The most abundant variety is cinnamon brown to resin brown in reflected light, and pale reddish brown to pinkish brown in transmitted light, with little or no pleochroism. The appearance of clinopyroxene in the crystalline rocks is generally stubby prismatic or anhedral, but sheaflike intergrowths containing feldspar are also present. Several crystals are strongly zoned from the center outward, as indicated by an increasing positive optic angle from near $0°$ to $50°$, together with an increasing refraction index and color intensity. The determined optical properties of the crystals appear to fit into the pigeonite-augite series and do not exclude titaniferous varieties. Cinnamon-brown pigeonite in samples of fines was identified by X-ray diffraction.

Olivine from Fo_{61} to Fo_{72} is a subordinate phenocrystic constituent of several of the finer crystalline rocks and occurs sporadically as crystal fragments in the breccias and dust. Olivine is a clear, pale greenish-yellow color in the crystalline rocks but may range from greenish yellow to honey yellow and orange yellow in the breccias and dust. The presence of olivine in the crystalline rocks was confirmed by X-ray diffraction methods.

Plagioclase is also abundant, but is generally found in a smaller amount than are the ferromagnesian minerals. Plagioclase is indicated by optical properties to be calcic, between An_{70} and An_{90}, with compositional zoning in several rocks. The appearance of plagioclase is commonly lath and plate shaped, with lamellar twinning parallel and transverse to the plates. The presence of calcic plagioclase in the crystalline rocks was confirmed by X-ray diffraction methods.

Ilmenite, as identified optically and by X-ray diffraction, is present in relatively large amounts in the crystalline rocks, where ilmenite occurs as plates and well-formed skeletal crystals. Ilmenite is also common in the breccias and dust as a constituent of the lithic fragments and as isolated crystal fragments.

Low cristobalite is present as thin, clear coatings and anhedral crystals. The cristobalite occurs in cavities and fills interstices between plagioclase plates in several of the coarser crystalline rocks. It is characterized by a crackly surface and complex twinning. Unidentified yellow transparent crystals occur in cavities interstitial to the plagioclase crystals. The mineral is a characteristic accessory in several of the more coarsely crystalline rocks.

Troilite (tentative identification) occurs in small amounts in several of the coarser crystalline rocks as rounded masses in interstices between plagioclase, clinopyroxene, or ilmenite. Native iron (tentative identification) occurs as scattered blebs up to 10 μm in diameter within the troilite masses. Several other accessory minerals are present in crystalline rocks but have not yet been identified.

Chemistry

Chemical analyses of the rock samples were made primarily by optical spectrographic methods operated inside the biological barrier. An instrument with a dispersion of 5.2 Å/mm was used. Three separate techniques were used to determine the chemical composition of the rocks:

(1) Determination of silicon, aluminum, iron, magnesium, sodium, potassium, calcium, titanium, manganese, and chromium, using strontium as the internal standard

(2) Determination of iron, magnesium, titanium, manganese, chromium, zirconium, nickel, cobalt, scandium, vanadium, barium, and strontium, using palladium as the internal standard

(3) Determination of lithium, rubidium, cesium, lead, copper, and other volatile elements, using sodium as the internal standard

The precision of the determinations was ±10 percent of the amount of an element present in a rock. Accuracy of the results was controlled by use of the international rock standard samples (G-1, W-1, Sy-1, BCR-1, DTS-1, PCC-1, G-2, AGV-1, and GSP-1) for calibration. Analyzed terrestrial basaltic rocks (Hawaii and Galapagos), chondritic meteorites (Forest City and Leedey), and achondrites (Sioux County and Johnstown) provided additional calibration points. All spectrographic line identifications were checked against individual-element spectra, in addition to the Massachusetts Institute of Technology wavelength tables and the U. S. Bureau of Standards tables.

The spectrographic plates were examined to establish the presence or absence of all elements that contained spectral lines in the wavelength regions covered. A list of elements that were not detected in the rock samples is given in table 5-I.

TABLE 5-I. *Elements undetected in rock samples*

Element	Detection limits, ppm	Element	Detection limits, ppm
Cesium	1	Gold	10
Beryllium	3	Zinc	30
Lanthanum	30	Cadmium	10
Neodymium	30	Mercury	100
Hafnium	50	Boron	5
Niobium	50	Gadolinium	10
Tantalum	100	Indium	1
Molybdenum	5	Thallium	1
Tungsten	20	Germanium	5
Rhenium	100	Tin	10
Rubidium	10	Lead	2
Rhodium	10	Arsenic	100
Palladium	10	Antimony	20
Iridium	50	Bismuth	20
Platinum	50	Tellurium	100

Line interferences caused by titanium, chromium, and zirconium lines were checked for all analytical lines. Wavelengths of the several principal lines were checked for those elements that were not detected in the rock samples, because line interferences caused by other elements were visible on some lines.

Following sterilization procedures, three rock samples were taken from behind the biological barrier and analyzed by atomic absorption procedures for the presence of iron, magnesium, calcium, titanium, sodium, and potassium. The rocks were also analyzed by a colorimetric procedure for the presence of silicon.

Sample weights were small, typically from 10 to 50 mg. Larger weights (150 mg) were available from those samples taken from behind the biological barrier, notably the biopool sample (LRL specimen 10054), which represented a 500-g sample. A total of 23 analyses were made on the rock samples. Analytical data resulting from these analyses and for 12 typical lunar samples are given in tables 5-II and 5-III.

The rock samples were apparently kept free from inorganic contamination from either the rock box or the LM. Niobium, present in the skirt of the descent engine exhaust (88 percent), was not detected (detection limit 50 ppm) and indium, which forms the seal of the rock box, was not present (detection limit 1 ppm).

Differences among the various rock samples were less apparent than the overall similarity in chemical composition (e.g., no samples contained less than 5 percent titanium dioxide). In detail, significant variations were shown in many of the trace elements, nickel, zirconium, rubidium, and potassium.

The major constituents of the rock samples are silicon, aluminum, titanium, iron, calcium, and magnesium. Sodium, chromium, manganese, potassium, and zirconium are minor constituents ranging from a few hundred parts per million to half a percent by weight.

The most striking features of the rock-sample chemical compositions in comparison with terrestrial rocks, meteorites, or cosmic abundance estimates are the high concentrations of titanium, zirconium, and yttrium. In comparison with chondritic meteorites, the rock samples contain lower concentrations of iron and magnesium and higher concentrations of calcium and aluminum.

Among the trace and minor constituents, zirconium, strontium, barium, yttrium, and ytter-

TABLE 5-II. *Elements detected in rock samples*

[Elemental abundances]

Element	Type A rocks (vesicular)				Type B rocks (crystalline)				Type C rocks (breccias)		Type D fine material	Bulk biopool sample
	10022	10072	10057	10020	10017	10058	10045	10050	10021	10061	10037	10054
Rubidium[a]	(b)	6.5	6.0	1.5	6.0	1.6	1.9	0.8	(b)	3.1	2.2	2
Barium[a]	100	130	180	50	120	85	115	60	105	90	68	65
Potassium[c]	.17	.17	.15	.053	.18	.09	.084	.053	.12	.15	.10	.11
Strontium[a]	110	55	230	85	55	190	60	140	150	60	90	140
Calcium[c]	6.4	6.8	7.1	7.1	7.1	7.5	7.1	7.1	7.9	7.9	8.6	8.3
Sodium[c]	.30	.44	.40	.44	.48	.41	.38	.38	.15	.37	.40	.38
Ytterbium[a]	7	2	6	2.5	(b)	5	1.3	2.7	4.5	1.8	2.5	2.5
Yttrium[a]	230	210	310	185	310	230	100	130	300	115	130	200
Zirconium[a]	1000	850	>2000	980	1250	250	700	700	1500	400	400	500
Chromium[a]	2800	4700	6500	2100	4600	3700	3500	4800	2500	3000	2500	2800
Vanadium[a]	36	30	40	20	30	32	40	80	22	32	42	30
Scandium[a]	110	45	110	110	55	130	90	170	68	55	55	60
Titanium[c]	6.6	6.0	7.5	7.2	6.6	5.4	4.8	5.4	5.2	5.4	4.2	4.2
Nickel[a]	320	(d)	25	(d)	(d)	(d)	(d)	55	215	235	250	120
Cobalt[a]	15	12	22	3	10	7	7	10	13	12	18	11
Copper[a]	(b)	5	(b)	4.5	3	(b)	6	10	(b)	8	(b)	(b)
Iron[c]	16	13	15.5	14	14.7	13	14	15.5	14.8	12.4	12.4	12.1
Manganese[a]	2000	2800	3800	2460	2700	4300	2100	3900	1700	2400	1750	2600
Magnesium[c]	3.9	4.8	5.7	4.8	5.1	3.9	4.2	6.0	4.5	5.4	4.8	4.6
Lithium[a]	11.5	14	22	15	25	19	15	10.5	(b)	12.5	15	15
Gallium[a]	(b)	(b)	(b)	5	(b)	(b)	4	8	(b)	(b)	(b)	(b)
Aluminum[c]	4.1	4.8	5.8	5.8	5.3	6.9	6.9	5.8	5.8	5.8	6.9	6.9
Silicon[c]	20	21	16.8	17.8	18.7	20	19.6	17.8	20	18.7	20	19.6

[a] In parts per million.
[b] No data.
[c] In percent.
[d] Not detected.

TABLE 5-III. *Elements detected in rock samples*

[Abundances expressed as weight, in percent oxides]

Oxide	Type A rocks (vesicular)				Type B rocks (crystalline)				Type C rocks (breccias)		Type D fine material	Bulk hopper sample
	10022	10072	10057	10020	10017	10058	10045	10050	10021	10061	10087	10084
SiO_2	43	45	36	38	40	43	42	38	43	40	43	42
Al_2O_3	7.7	9	11	11	10	13	13	11	11	12	13	13
TiO_2	11	10	12.5	12	11	9	8	9	8.6	10	7	7.0
FeO	21	17	20	18	19	17	18	20	19	16	16	15.6
MgO	6.5	8	9.5	8	8.5	6.5	7	10	7.4	9	8	7.6
CaO	9.0	9.5	10	10	10	10.5	10	10	11	11	12	11.6
Na_2O	.40	.60	.54	.59	.65	.56	.51	.51	.20	.48	.54	.50
K_2O	.21	.20	.18	.064	.22	.11	.10	.064	.15	.17	.12	.14
MnO	.28	.36	.49	.32	.35	.55	.27	.50	.22	.41	.23	.34
Cr_2O_3	.41	.69	.95	.31	.67	.54	.51	.70	.37	.69	.37	.41
ZrO_2	.14	.11	>.27	.13	.19	.03	.095	.095	.20	.04	.05	.07
NiO	.04	(a)	(a)	(a)	(a)	(a)	(a)	.007	.03	.04	.03	.015
Total	99.0	100.5	101.4	97.8	100.5	100.8	99.5	99.9	99.8	99.8	100.3	98.2

ª No data.

bium are enriched by one to two orders of magnitude compared to chondritic abundances. Potassium and rubidium are present in similar amounts, and nickel and cobalt are depleted by large factors compared to chondrites. Nickel was not detected (less than 1 ppm) in several rock samples. However, the iron content remains high in these samples, and the iron-nickel ratios are high. The concentration of zirconium is equaled only by alkali-rich rocks in the crust of the Earth, such as nepheline syenites.

Rubidium-strontium ratios are low, approaching those of oceanic basaltic rocks on the Earth. Barium is relatively abundant and, like lithium, resembles the concentration levels of continental basaltic rocks (approximately 10 ppm). Chromium is notably abundant as a minor constituent, and scandium is also present in larger amounts than are generally found in terrestrial basalts. Conversely, vanadium is less abundant than either chromium or scandium.

The volatile elements (lead, bismuth, thallium, etc.) are generally below the detection limits of the spectrographic techniques employed. The platinum-group elements, silver and gold, were not detected. Thus, numerous differences in detail exist from terrestrial or meteoritic samples previously available for analysis.

Rare-Gas Analytic Results

The gas analyses were obtained by using a 6-in., 60° magnetic deflection mass spectrometer. The spectrometer is operated with a sensitivity of approximately 2×10^{-14} cc/mV for helium, neon, and argon and approximately 3×10^{-13} cc/mV for krypton and xenon. Rock samples were prepared under liquid nitrogen and air conditions. Chips of rock were taken, weighed, wrapped in aluminum foil, and sterilized by heating from 125° to 150° C for a 5- to 24-hr period. The effect of the sterilization heatings caused a release of less than 1 percent of the gases present.

The rock samples were melted by radiofrequency induction heating in a molybdenum crucible. Vacuum conditions were maintained by a water-cooled glass furnace with low helium-diffusion characteristics. The gases released were chemically purified with a hot titanium getter. The heavier noble gases were condensed and introduced into the mass spectrometer for measurement under static vacuum conditions in three fractions: (1) helium and neon, (2) argon and krypton, and (3) xenon.

In general, after each rock sample was run, a second heating was performed to insure that all the gas had been released from the rock and to serve as a blank correction. The entire procedure was standardized by introducing a calibrated amount of the different rare gases into the mass spectrometer.

The rare gases in the lunar samples show three general patterns, which correspond to the three rock types (breccias, igneous rocks, and fines). Generally, the breccias and fines show extremely large concentrations of rare gases. Large amounts of noble gases are found in both surface and subsurface breccia material. From the large amounts of noble gases present and the isotopic ratios, it is almost certain that the predominant source for these gases is the solar wind. The only other possible major source would be rare gases condensed in lunar materials early in the history of the solar system. The igneous rocks contain less of the rare gases than do the other rocks. This phenomenon can be interpreted either as the result of a loss of rare gases during formation, or as a feature of igneous material representing subsurface lunar material that has not been exposed to the solar wind except as a solid. Temperature-release studies on both the fines and the breccia reveal that the noble gases are tightly bound and thus do not occur as simple surface-absorbed gases.

The breccias appear to have been formed from the fines near the surface of the Moon as a result of some compacting process, possibly shock. This conclusion is supported by petrological evidence. The solar-wind noble gases may have been driven into the separate grains as a result of these shock events.

The results shown in table 5–IV indicate large amounts of rare gases of solar composition. No isotopic ratios have been measured previously from the Sun; however, from theoretical considerations, it is generally accepted that nuclear processes occur that cause the solar neon 20/22 ratio to be larger than that of the atmosphere of the Earth. This enrichment is clearly present in all the samples measured. In addition, the helium on the surface of the Sun should be close to the original helium content of the solar system. The ^{4}He/^{3}He ratio of approximately 2600 agrees well with previous theoretical estimates. The low value of the ^{40}Ar/^{36}Ar ratio is exactly what would be expected for solar isotopic composition. The relative elemental ratios for the noble gases are those that would be expected in the Sun if considerable helium and some neon are assumed to be lost from the lunar surface during the hot lunar day.

The isotopic ratios for xenon, presumed to be of solar origin, in the fines and in a breccia are presented in table 5–V. Isotopically, the xenon in the two samples is identical. The isotopic pattern closely resembles that of trapped xenon found in carbonaceous chondrites, except for a slightly larger amount of the lighter xenon isotopes (presumably from spallation reactions) and except for a deficiency of ^{134}Xe or ^{136}Xe. The ^{129}Xe/^{132}Xe ratio in the measured material is essentially identical to that in carbonaceous chondrites. Krypton isotope data for the fine lunar material are less precise, but resemble krypton content found in carbonaceous chondrites.

Several of the crystalline rocks measured con-

TABLE 5–IV. *Rare-gas analysis*

Sample description	Helium		Neon			Argon			Krypton	Xenon
	10^{-8} cc/g ^{4}He	4/3	10^{-8} cc/g ^{20}Ne	20/22	22/21	10^{-8} cc/g ^{40}A	40/36	36/38	10^{-8} cc/g	10^{-8} cc/g
Lunar soil:										
Sample A	10 900 000	2500	200 000	13	31	39 000	1.1	5.3	38	38
Sample B	19 000 000	2500	310 000	13	30	42 000	1.2	5.4	36	16
Typical agglomerate rock:										
Sample A	15 100 000	2900	320 000	13	29	150 000	2.3	5.2	73	46
Sample B	16 000 000	2700	230 000	13	29	110 000	2.2	5.2	49	42
Typical crystalline rock:										
Sample A	63 000	180	210	3.1	1.3	5 700	96	1.2	.34	.85
Sample B	28 000	270	200	7.0	2.3	1 600	42	2.4	.19	.16

TABLE 5–V. $^iXe/^{132}Xe$ ratios

Sample	^{124}Xe	^{126}Xe	^{128}Xe	^{129}Xe	^{130}Xe	^{131}Xe	^{134}Xe	^{136}Xe
Fines	0.0062	0.0071	0.086	1.07	0.165	0.829	0.373	0.306
Breccia	.0052	.0057	.084	1.07	.164	.820	.37	.304

tain radiogenic ^{40}Ar and spallation-produced noble gases which, when coupled with potassium contents, enable potassium argon ages and cosmic-ray radiation ages to be estimated. Seven rocks have yielded potassium argon ages, all of which are consistent with a value of $3.0 \pm 0.7 \times 10^9$ yr. Radiation ages show a wider variance among different rocks, ranging from approximately 10^7 yr to approximately 16×10^7 yr. The relative amounts of spallation-produced helium, neon, and argon differ from those in most stone meteorites and probably reflect the unique chemistry of the lunar material.

Gamma-Ray Spectrometry

The Radiation Counting Laboratory (RCL) received eight lunar samples for preliminary study by nondestructive gamma-ray analysis. These samples included one rock from the contingency sample, five rocks from the documented-sample box, and a sample of fines and one rock from the bulk-sample box.

Because of the complex operations involved in handling lunar material, preparation of samples for analysis in the RCL was slow, and analysis of the first RCL sample could not begin until July 29, 1969. As a result, radioactive species with half lives of less than a few days were not detectable. In addition, the intense interferences from the gamma-ray spectra of the thorium and uranium decay series in the samples made detection of weak gamma-ray components very difficult.

The low-background gamma-ray spectrometers used in these studies were located 15 m below the ground in a room supplied with radon-free air and shielded by 0.9 m of compacted dunite inside a welded steel liner of low radioactivity. The principal detector system consisted of two 23-cm-diameter, 13-cm-long NaI(Tl) detectors at 180° with the sample between them. Data were recorded in the singles and coincidence modes. Response of the detectors was enhanced by use of a surrounding anticoincidence mantle. Background was reduced further by surrounding the detectors and the inner mantle with a thick lead shield, the exterior of which was covered by a thin meson-sensitive anticoincidence mantle. Spectra were also recorded with a large-volume Ge(Li) detector, which was located inside a lead shield 10 cm thick. However, all data in this report were obtained from the NaI(Tl) scintillation spectrometer. With the exception of the preamplifiers, all ancillary electronic instrumentation was located in an adjacent control room.

For the preliminary investigation, the equipment was calibrated with the aid of a series of standards prepared by dispersing known amounts of radioactivity into quantities of electrolytically reduced iron powder. The weights of the cylindrical dispersed samples, which had a density of 3.4 g/cc, overlapped the range of anticipated lunar-sample weights. Lunar samples were transferred to the RCL for analysis in a variety of nonstandard containers. The material of the container affects the gamma-ray response through gamma-ray scattering and absorption. Where feasible, a library of standard spectra was recorded with dispersed sources inside the type of container used for the actual lunar sample. However, this procedure could not be followed in every instance. The activity of each radionuclide was determined by using the method of least squares to resolve the gamma-ray spectra.

Until it is possible to standardize the equipment with phantoms that have the same shape and density and that contain known amounts of the radionuclides of interest, the results of the preliminary studies have been assigned large errors. Some samples were processed for analysis without being weighed. In such cases, the sample weight was estimated by correcting the weight of the sample package for the weight of the container, packing material, and biological containment bags.

The results of the preliminary studies are

summarized in table 5-VI. Twelve radioactive species were identified; some identifications were only tentative. The nuclides of shortest half life that were characterized were ^{52}Mn (half life, 5.7 days) and ^{48}V (half life, 16.1 days). It is important to note that the concentrations listed in table 5-VI represent averages for the entire sample (usually a rock) and, therefore, are not subject to sampling errors if the rock is inhomogeneous.

The amount of potassium present is variable and ranges over the average potassium concentration (0.085 weight percent) for chondrites. Conversely, the uranium and thorium concentrations are near the values for terrestrial basalts, and the thorium-to-uranium ratio is approximately 4.1, with little variation. Despite variation of a factor of 5.6 in the potassium concentration, the potassium-to-uranium ratio determined for lunar surface material (2400 to 3100) is remarkably constant and much lower than similar ratios for terrestrial rocks or meteorites.

Yields of cosmogenic ^{26}Al are generally high, and the ^{22}Na/^{26}Al ratio is considerably below unity in both the lunar rocks and the soil. The high yields of ^{26}Al may be understood in terms of the chemical composition of the lunar surface material, if the samples in question have been exposed to cosmic-ray bombardment for several half lives (0.74×10^6 years) of ^{26}Al so that the production of ^{26}Al could reach saturation. Such reasoning suggests that the material analyzed in these studies has been exposed to cosmic radiation for at least several million years, which is in agreement with the rare-gas analyses.

Magnetic Measurements

Preliminary magnetic measurements have been completed on 31 samples of crystalline rock and breccia, using a triaxial fluxgate gradiometer in the vacuum chamber and class III cabinets. The level of significance of the results from various samples ranged from 1×10^{-5} to 1×10^{-2} emu/g. The significance level is a function of sample weight, sample shape, intensity of magnetization, and the highly variable magnetic interference generated by the cabinets around the sample.

Of the 13 samples of crystalline rock measured, seven gave no measurable magnetic re-

Table 5-VI. *Summary of gamma-ray analyses of lunar samples* [a]

	10057	10072	10003	10017	10018	10019	10021	10022
Weight, g	897	[b]399	[b]213	971	[b]213	[b]245	[b]216	302
Classification	A	A	B	B	C	C	C	D
Potassium,[c] weight percent	0.242±0.036	0.232±0.035	0.050±0.008	0.227±0.0034	0.144±0.022	0.12±0.02	0.120±0.018	0.11±0.02
Thorium, ppm	3.4±0.7	2.9±0.4	0.85±0.14	2.9±0.4	2.3±0.3	1.9±0.3	1.8±0.3	1.8±0.3
Uranium, ppm	.78±0.16	.75±0.11	.20±0.03	.70±0.10	.60±0.09	.43±0.08	.38±0.06	.46±0.10
^{26}Al, dpm/kg	77±16	70±15	68±14	66±13	100±20	98±20	81±16	97±19
^{22}Na, dpm/kg	44±9	42±9	41±8	34±7	55±11	47±10	41±8	±9
^{44}Ti, dpm/kg	[d]TI	[d]TI	[d]TI	—	—	—	—	TI
^{46}Sc, dpm/kg	10±3	13±4	13±3	11±3	13±4	10±4	10±4	9±3
^{48}V, dpm/kg	—	—	[d]TI	[d]TI	—	—	—	—
^{52}Mn, dpm/kg	—	—	39±18	—	—	—	—	—
^{54}Mn, dpm/kg	40±13	20±8	26±5	38±13	28±14	27±10	30±20	28±9
^{56}Co, dpm/kg	30±12	30±10	38±6	18±6	33±11	35±11	15±7	27±10
^{7}Be, dpm/kg	TI	TI	TI	TI	—	—	38±13	TI

[a] Values for short-lived nuclides have been corrected for decay to 00:00 hr c.d.t., July 21, 1969.
[b] Weight uncertain, see text.
[c] Potassium determined by assaying ^{40}K and assuming terrestrial isotopic ratios for potassium.
[d] Tentatively identified.

sponse. In six samples, induced magnetization of approximately 1×10^{-4} emu/g in fields ranging from 0.3 to 1.9 Oe and with uncertainties from 25 to 50 percent was detected. One of these six samples also had a barely detectable remnant magnetization, approximately 3×10^{-4} emu/g with an uncertainty of approximately 50 percent.

The breccias are approximately 10 times as magnetic as the crystalline rocks. Eighteen breccias were measured, five of which gave no measurable magnetic response. Seven of the breccia samples showed only induced magnetization, whereas both induced and remnant magnetization was detected in five samples. Intensities of induced magnetization in various breccia samples ranged from 1×10^{-12} to 1×10^{-4} emu/g in fields ranging from 0.3 to 0.9 Oe and with uncertainties from 10 to 50 percent. The intensity of remnant magnetization in the five samples was approximately 1×10^{-3} emu/g with uncertainties from 20 to 50 percent. The ratio of remnant to induced magnetization for these five samples varied between 0.4 and 0.9 Oe. Several portions of fines from the contingency and bulk samples were also measured; the specific susceptibilities of these fines ranged from 1×10^{-2} to 4×10^{-2} emu/g, which is roughly 10 times greater than the specific susceptibilities of the breccias.

Organic Chemical Investigations

A simple survey method and a more elaborate and specific technique have been used to estimate the abundance levels of organic matter in the lunar samples. The survey method and the specific technique are (1) a pyrolysis-flame ionization detector capable of yielding an estimate of the total organic matter in the sample, regardless of type and origin, and (2) a computer-coupled high-sensitivity mass spectrometer capable of yielding detailed mass-spectrometric data on the volatile and/or pyrolyzable organic matter as a function of sample temperature. From the mass-spectrometric data (combined with the vaporization characteristics), an assessment may be made of the relative contributions of terrestrial contaminants versus the possible indigenous lunar organic matter.

Pyrolysis-Flame Ionization Detector Determination of Total Organic Carbon

Fourteen samples of lunar material were analyzed for total organic carbon content as part of the preliminary examination at the LRL. These samples included both fines and different types of rocks from the contingency pouch and the bulk and documented boxes.

The technique used in this determination is based on the pyrolysis of small portions (10 to 40 mg) at 800° C in a Nichrome tube under a flowing stream of hydrogen and helium, with subsequent detection of the volatile products by a hydrogen flame ionization detector. The combustion of the organic volatiles produces metastable oxygenated species in the flame zone which undergo chemi-ionization to yield an ion current proportional to the number of carbon atoms entering the flame. After amplification and digital integration, the resultant single peak area is compared with known standards on a log-log least-squares calibration plot to yield the organic carbon content in parts per million.

The detector response is highly specific to organic compounds and does not result from such inorganic gases as carbon dioxide, carbon monoxide, water, nitrogen, oxygen, sulfur oxides, nitrogen oxides, hydrogen halides, etc. Graphite and diamond are not detected by this technique. Terrestrial soils give values of 30 to 500 ppm from dry desert areas and values of 500 to 20 000 ppm from agricultural areas. Ancient sediment rocks can yield values as low as 50 ppm when most of the organic matter has been converted into kerogenic material. By this method, a portion of the recently fallen Puebito de Allende meteorite was found to contain 37 ppm of organic carbon, and a sample of volcanic bomb was found to have less than 1 ppm.

The apparatus for the determination is shown in schematic form in figures 5–8 and 5–9 and is shown in figure 5–10 as installed in the Class III cabinetry in the Physical Chemistry Test Laboratory of the LRL.

The fines from the contingency sample were sampled 1 day after the plastic pouch had been opened and the rocks removed with metallic implements. This sample (sample 10010,5), which indicated 16 ± 7 ppm of organic matter,

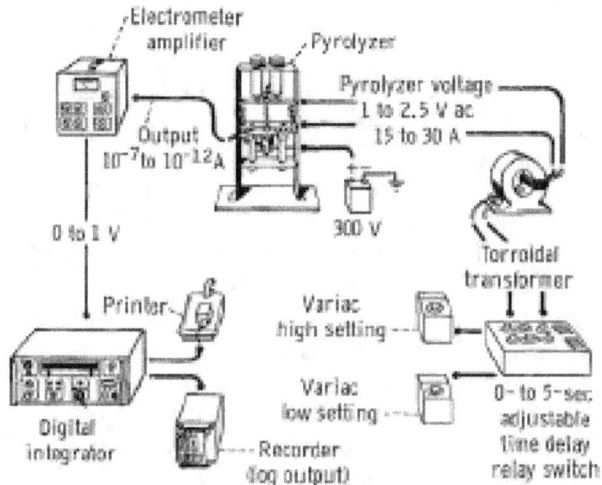

FIGURE 5-8. — Organic carbon electronics schematic.

FIGURE 5-10. — Biological enclosure, organic carbon determination.

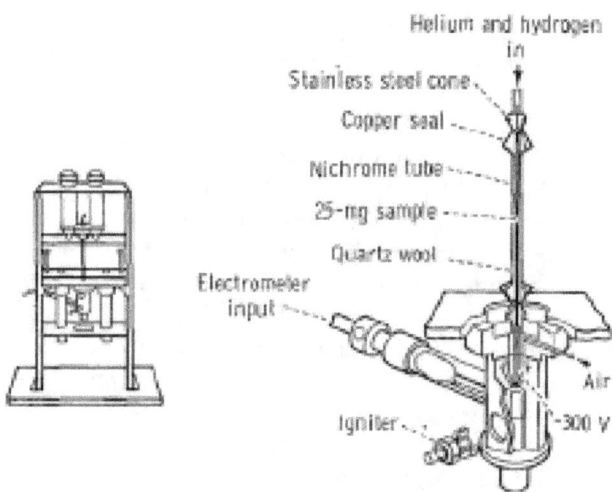

FIGURE 5-9. — Organic carbon pyrolyzer-flame ionization detector.

was exposed to the spacecraft and terrestrial atmosphere during the return trip but was handled carefully with respect to organic contamination. Therefore, it was concluded that this sample was not heavily contaminated. Eleven days later, after many intrusions into the sample for other experiments, another portion of these fines was analyzed, showing 48 ± 18 ppm of organic matter. Microscopic examination of these fines showed white fibers, which were judged to be cellulose contamination.

Three rocks were analyzed from the contingency sample. An aggregate (sample 10021,21) was chipped after extensive geologic examination (frequent rubber-glove contact) to give 44 ppm of organic matter. The crystalline rock (sample 10024,4) was analyzed after only brief geologic examination to give 17 ppm of organic matter. Both these rocks had been stored in plastic boxes. A third contingency rock (sample 10032,1), previously unexamined, was carefully chipped under clean conditions and showed 10 ppm. These results from the contingency samples show that even moderate handling can easily introduce low amounts of organic-matter contamination. The primary contamination sources appear to be contact with rubber gloves, plastic materials, fiber brushes and tissues, and air-Freon spray for dust removal. However, because of the relatively simple processing procedure used with the contingency sample, fairly clean samples can be obtained with adequate precaution.

The documented box was opened in the F-201 vacuum chamber, which was known to contain significant amounts of pump oil. In addition, the relatively cumbersome hardsuit gloves made sample processing more complex and subject to contamination.

The fines (sample 10015,10) from the gas reaction cell were analyzed after exposure to nitrogen, carbon dioxide, and oxygen and were found to give 126 ppm. White filaments were observed in this sample.

Three aggregate rock samples were taken from chipping operations in the vacuum cham-

TABLE 5-VII. Apollo 11 lunar sample analysis for total organic carbon

Sample origin	Sample	Type	Net organic carbon, ppm	Control blank	No. of runs	Cabinet processed	Comments
Contingency	10010,5	Fines	16	7	2	P-chem[a]	Sampled 1 day after pouch opened
	10010,26	Fines	48	18	3	P-chem[a]	Sampled 11 days later, fibers noted
	10021,21	Aggregate	44	31	3	P-chem[a]	Plastic box storage, extensive handling
	10024,4	Crystalline	17	6	2	P-chem[a]	Plastic box storage, moderate handling
	10032,1	Crystalline	10	9	1	P-chem[a]	Minimal handling
Documented	10015,10	Fines	126	64	1	F-201[b]	Freon gas reaction cell heavily handled, fibers noted
	10061,4	Aggregate	62	19	2	F-201[b]	No special handling, clean bench
	10064,4	Aggregate	96	30	2	F-201[b]	No special handling, clean bench
	10068,3	Aggregate	39	18	1	F-201[b]	Interior chip, special handling, clean bench
Bulk	10002,73	Fines	18	8	2	Bio Prep[c]	Sieved, special handling for minimum contamination
	10046,3	Aggregate	80	26	2	Bio Prep[c]	P-chem chip, routine handling
	10048,3	Aggregate	88	43	2	Bio Prep[c]	P-chem chip, routine handling
	10049,7	Crystalline	54	23	1	Bio Prep[c]	P-chem chip, routine handling
	10050,3	Crystalline	28	19	3	Bio Prep[c]	P-chem chip, routine handling
	10056,12	Aggregate	10	4	2	Bio Prep[c]	Specially chipped and handled for minimum organic contamination

[a] In the Physical-Chemical Test Laboratory of the LRL.
[b] Indicates the vacuum chamber of the LRL.
[c] In the Biological Preparation Laboratory of the LRL.

ber after the glove implosion and subsequent heat sterilization. These chips were processed outside the F-201 vacuum chamber in a sealed can, and portions were taken on a clean bench with organic cleanliness. Two chips were designated for physical-chemical examination and were handled an unknown amount by the glove operator. These samples (10061 and 10064) indicated 62 and 96 ppm. The third chip (sample 10068) was an interior-surface fragment sampled in the F-201 vacuum chamber with metal tongs, resulting in 39 ppm, which was distinctly lower than results from the other two fragments from the vacuum chamber. This result shows the amount of contamination from F-201 handling to be from 30 to 40 ppm from the gaseous components and an additional 20 to 60 ppm from other handling.

The bulk box was processed in the biological preparation cabinetry under dry nitrogen, but the samples in this box were inadvertently exposed to Freon and ethylene oxide during the initial sterilization of the exterior of the box. The rock chips from this box gave results between 28 and 88 ppm when handled routinely. Special samples (processed for organic analysis) showed marked improvement. A chip from sample 10056 gave 10 ppm, and the bulk fines (after sieving) indicated 18 ppm. The major sources of contamination probably result from handling and the rigorous sterilization procedures necessary for introduction or removal of any materials from the biological preparation cabinets.

The results of these tests are summarized in table 5-VII. The data indicate that the indigenous organic content of the lunar samples is below 10 ppm. Contamination from handling, although low, can easily mask the native organic matter. The samples returned from future Apollo missions should give a better estimate of the organic levels if adequate precautions are taken to avoid organic contamination.

Mass Spectrometry

The mass-spectrometric investigation involved examination of the lunar sample by repetitive mass-spectrometric monitoring of the vapor phase evolved by heating the lunar sample from ambient temperature to approximately 500° C. A mass spectrometer was operated on line with a Scientific Data Systems Sigma 2 computer. The mass spectrometer was designed to meet the LRL biological-containment requirements, which necessitate a sterilizable inlet system, and the mass spectrometer was equipped with a high-efficiency ion source and a high-gain electron multiplier. The online computer permitted continuous control of the scanning circuit, recording of the mass spectra, and calibration of the mass scale, spectrum normalization, and visual (cathode-ray tube) display of the spectrum during each scan cycle.

The lunar material (approximately 50 mg) was placed in a nickel capsule, sealed by a cold weld, and transferred to the instrument inlet system. The inlet system allowed insertion of these capsules into an oven (at approximately 500° C) connected to the ionization chamber by a heated 8-cm quartz tube.

To calibrate the system, three samples of n-tetracosane on clean quartz (at concentrations of 1, 2, and 3 ppm) were analyzed with the mass spectrometer/computer system to determine the total ionization recorded versus the amount of organic material present. (Concentration levels reported refer to an equivalent signal of n-tetracosane.) Such concentrations are relatively large in terms of instrument sensitivity.

The lunar material studied came from three distinct sample types: (1) the core tubes; (2) fines (type D) from the bulk box; and (3) interior and exterior chips, fine- and medium-grained (types A and B) and breccia (type C) rocks that were returned in both boxes. Control samples of outgassed clean sand (Ottawa sand), which were exposed to the sample cabinet atmosphere for a length of time similar to the lunar samples' exposure, were analyzed to estimate the amount of organic contamination introduced by exposure to the atmosphere.

The levels of organic matter observed from this experiment, whether indigenous or of terrestrial origin (contamination), were extremely low (0.2 to 5.0 ppm). Because of this low level, the quantity of organic matter determined can vary widely with the slightest amount of organic contaminant from anywhere in the entire sample-handling and storage procedure between the Moon and the mass spectrometer. The analyses of blanks and controls from all phases of lunar-sample-handling and storage procedures, usually using clean quartz, indicate that most of the organic material observed in the lunar samples is caused by organic matter contamination.

Data on a sample obtained from the core tube and on a sample of fines from the bulk box may indicate the presence of an extremely small amount of pyrolyzable organic matter. This assumption is based on the observation that the abundance of ions at m/e 78 and 91 increased in the spectra obtained later in the heating cycle of these samples. These may be interpreted as aromatic hydrocarbon ions that are indicative of the products of the pyrolysis of indigenous organic matter.

In conclusion, the maximum levels of indigenous organic matter indicated by the observations described in this report appear to be considerably less than 1 ppm.

Biology

The preliminary examination of samples has included microscopic studies to find any living, previously living, or fossil material, if possible. No evidence of any such material has been found.

Approximately 700 g of fines and rock chips have been subjected to an extensive biological protocol. A large variety of biological systems was challenged with lunar material. The biological systems included the following:

(1) Germ-free mice
(2) Fish
(3) Quail
(4) Shrimp, oysters, and other invertebrates
(5) Several lines of tissue cultures
(6) Several varieties of insects
(7) A considerable number of plants
(8) Lower animals, including paramecia

No evidence of pathogenicity had been observed when this report was written (August 23, 1969).

Conclusions

The major findings of this preliminary examination of the lunar samples are as follows:
(1) The fabric and mineralogy of the rocks allow the rocks to be divided into two genetic

groups: (a) fine- and medium-grained crystalline rocks of igneous origin, probably originally deposited as lava flows, dismembered, and redeposited as impact debris and (b) breccias of complex history.

(2) The crystalline rocks are different from any terrestrial rock and from meteorites, as shown by the modal mineralogy and bulk chemistry.

(3) Erosion has occurred on the lunar surface, as indicated by the rounding on most rocks and by the evidence of exposure to a process that gives the rocks a surface appearance similar to sandblasted rocks. No evidence exists of erosion by surface water.

(4) The probable presence of the assemblage iron-troilite-ilmenite and the absence of any hydrated phase suggest that the crystalline rocks were formed under extremely low partial pressures of oxygen, water, and sulfur (in the range of those in equilibrium with most meteorites).

(5) The absence of secondary hydrated minerals suggests that there has been no surface water at Tranquility Base at any time since the rocks were exposed.

(6) Evidence of shock or impact metamorphism is common in the rocks and fines.

(7) All the rocks display glass-lined surface pits that may have been caused by the impact of small particles.

(8) The fine material and the breccia contain large amounts of all noble gases with elemental and isotopic abundances that almost certainly were derived from the solar wind. The fact that interior samples of the breccias contain these gases implies that the breccias were formed at the lunar surface from material previously exposed to the solar wind.

(9) The $^{40}K/^{40}Ar$ measurements on igneous rock indicate that those rocks crystallized 3×10^9 to 4×10^9 yr ago. Cosmic-ray-produced nuclides indicate the rocks have been within 1 m of the surface for periods of 20×10^6 to 160×10^6 yr.

(10) The level of indigenous volatilizable and/or pyrolyzable organic material appears to be extremely low (i.e., much less than 1 ppm).

(11) The chemical analyses of 23 lunar samples show that all rocks and fines are generally similar chemically.

(12) The elemental constituents of lunar samples are the same as those found in terrestrial igneous rocks and meteorites. However, several significant differences in composition occur: some refractory elements (e.g., titanium and zirconium) are notably enriched and the alkalis and some volatile elements are depleted.

(13) Elements that are enriched in iron meteorites (i.e., nickel, cobalt, and the platinum group) were not observed, or such elements were low in abundance.

(14) Of 12 radioactive species identified, two were cosmogenic radionuclides of short half life, namely ^{52}Mn (half life 5.7 days) and ^{48}V (half life 16.1 days).

(15) Uranium and thorium concentrations lie near the typical values for terrestrial basalts; however, the potassium-to-uranium ratio determined for lunar surface material is much lower than such values determined for either terrestrial rocks or meteorites.

(16) The high observed ^{26}Al concentration is consistent with a long cosmic-ray exposure age inferred from the rare-gas analysis.

(17) No evidence of biological material has been found in the samples to date.

(18) The lunar soil at the LM landing site is predominantly fine grained, granular, slightly cohesive, and incompressible. The hardness increases considerably at a depth of 6 in. The soil is similar in appearance and behavior to the soil at the Surveyor landing sites.

Discussion

The data and descriptive information in this report were obtained to characterize the materials about to be distributed to principal investigators and their associates for specialized and detailed study. The usefulness of this information in the selection of material for particular experiments is well illustrated by the rare-gas data. It would have been impossible to select material suitable for a careful study of the cosmic-ray-produced isotopes of neon, argon, krypton, and xenon, or even for a straightforward potassium-argon age without the knowledge of the variations in rare-gas content shown in table 5-IV. Similarly, the organic matter analyses provide guidance in selecting material for further experiments and study of organic matter in lunar materials.

In spite of the limited and specific objectives, the preliminary examination has provided significant, and a few unexpected, results on long-recognized questions.

The existence of an unexplained erosion process on the lunar surface was clearly indicated by both the Ranger and the Lunar Orbiter photographs. These photographs frequently show very fresh block craters interspersed among smoothed craters. Photographs of individual rocks on Surveyor photographs give further evidence of rounding and abrasion of hard rocks on the lunar surface. The surface morphology, glass pits, and splashes seen on both hard and fragmental rocks suggest that samples are now available for detailed laboratory examination that may elucidate a widespread and important mechanism on the lunar surface. The evidence provided by the first examination of these rocks indicates that this process is unlike any process so far observed on Earth.

The chemical composition of the Tranquility Base fines and igneous rocks is unlike those of any known terrestrial rock or meteorite. The unique characteristic is the unusually high titanium, zirconium, yttrium, and chromium content in the lunar rocks, compared to other rocks with this approximate bulk composition. Also of great interest is the low sodium, potassium, and rubidium content. It is particularly significant that the unique composition is that of a silicate liquid. If this liquid has a volcanic origin, the unique composition implies either that the composition of the rock from which the liquid was derived differs significantly from the composition of the mantle of the Earth, or that the mechanism by which the liquid was formed and/or the fractional crystallization of this liquid on the Moon may also differ from terrestrial analogs. The nearly identical composition found for the fines, fragmental rocks, and igneous rocks suggests that the unique composition observed for the Tranquility Base materials is characteristic of that part of the Moon and not the result of a local isolated flow or intrusion.

Several specific geochemical observations on the igneous rocks can be made from the present data. The potassium-to-uranium ratios (2400 to 3200) of the lunar materials are unusually low, both when compared to chondritic meteorites (45 000) and when compared to most common Earth rocks (10 000). This ratio is not readily changed by terrestrial igneous processes (refs. 5-1 and 5-2). If this generalization is extended to lunar igneous processes, observed ratios can be inferred to be characteristic of the entire Moon. Similarly, the rubidium-to-strontium ratio is more like that of the Earth and achondritic meteorites than that of chondritic meteorites and the Sun. Both these chemical characteristics suggest that, relative to chondrites, the Moon, like the Earth, is depleted in alkali metals greater than sodium in atomic number (ref. 5-3).

Variations in other elemental ratios that are easily fractionated during igneous differentiation processes (e.g., nickel/magnesium and barium/strontium) suggest that chemical differentiation has occurred in the formation of the igneous rocks. The abundance of radioactive elements (potassium, uranium, and thorium) in the surface materials is much greater than that inferred (from thermal models of the Moon) for the mean radioactive element content of the Moon (ref. 5-4). This finding leads to the inference that the surface materials are chemically differentiated with respect to the entire Moon.

The unusually high abundance of high-atomic-number elements (iron and titanium) is clearly consistent with the unusually high densities (3.1 to 3.5 g/cc) reported previously. Both the Surveyor alpha back-scattering analyses (ref. 5-5) and optical studies (ref. 5-6) indicate that mare materials may have significantly higher iron contents than highland materials. Assuming the rather plausible generalization that the densities of Tranquility Base rocks are characteristic of other mare regions, one can infer that large areas of the lunar surface, particularly the mare regions, may be made of materials with densities in excess of the mean density of the Moon.

Perhaps the most exciting and profound observation made in the preliminary examination is the great age of the igneous rocks from this lunar region. The potassium argon age is both intrinsically and experimentally uncertain; nevertheless, it is clear that the crystallization of some Apollo 11 rocks may date back to times earlier than the oldest rocks found on Earth. It seems likely that if the age of the Apollo 11 rocks does not date back to the time of formation of the Moon, rocks from other regions of the Moon will.

References

5-1. BIRCH, A. F.: Differentiation of the Mantle. Geol. Soc. Am. Bull., vol. 69, no. 4, Apr. 1958, pp. 483-486.

5-2. WASSERBURG, G. J.; MACDONALD, G. J. F.; HOYLE, F.; and FOWLER, W. A.: Relative Contributions of Uranium, Thorium, and Potassium to Heat Production in the Earth. Science, vol. 143, no. 3605, 1964, pp. 465-467.

5-3. GAST, P. W.: History of the Earth's Crust. R. A. Phinney, ed., Princeton University Press, 1968, pp. 15-27.

5-4. ANDERSON, D. L.; and PHINNEY, R. A.: Early Thermal History of the Terrestrial Planets. In Mantles of the Earth and Terrestrial Planets, S. K. Runcorn, ed., Interscience Publishers (London), 1967, pp. 113-126.

5-5. TURKEVICH, A. L.; ANDERSON, W. A.; ET AL.: Surveyor Project Final Report, Part II. Science Results. NASA TR 32-1265, 1968.

5-6. ADAMS, J. B.: Lunar and Martian Surfaces: Petrologic Significance of Absorption Bands in the Near-Infrared. Science, vol. 159, no. 3822, 1968, pp. 1453-1455.

Bibliography

Space Science Board, National Academy of Sciences: Conference on Potential Hazards of Back Contamination from the Planets, July 29-30, 1964.

National Aeronautics and Space Administration; the Department of Agriculture; the Department of Health, Education, and Welfare; the Department of the Interior; and the National Academy of Sciences: Interagency Agreement on the Protection of the Earth's Biosphere from Lunar Sources of Contamination, August 24, 1967.

McLANE, J. C., JR.; KING, E. A., JR.; FLORY, D. A.; RICHARDSON, K. A.; DAWSON, J. P.; KEMMERER, W. W.; and WOOLEY, B. C.: The Lunar Receiving Laboratory. Science, vol. 155, 1967, pp. 525-529.

SHOEMAKER, E. M.; MORRIS, E. C.; BATSON, R. M.; HOLT, H. E.; LARSON, K. B.; MONTGOMERY, D. R.; RENNILSON, J. J.; and WHITAKER, E. A.: Jet Propulsion Laboratory Tech. Rept. 32-1265, 1969, p. 21.

ACKNOWLEDGMENTS

The following people contributed directly in obtaining the data and in the preparation of this report: D. H. Anderson, Manned Spacecraft Center (MSC); E. E. Anderson, Brown and Root-Northrop (BRN); Klaus Bieman, Massachusetts Institute of Technology (MIT); P. R. Bell, MSC; D. D. Bogard, MSC; Robin Brett, MSC; A. L. Burlingame, University of California at Berkeley; W. D. Carrier, MSC; E. C. T. Chao, United States Geological Survey (USGS); N. C. Costes, Marshall Space Flight Center; D. H. Dahlem, USGS; G. B. Dalrymple, USGS; R. Doell, USGS; G. C. Davis, Ames Research Center; J. S. Eldridge, Oak Ridge National Laboratory (ORNL); M. S. Favaro, United States Department of Agriculture (USDA); D. A. Flory, MSC; Clifford Frondel, Harvard University; R. Fryxell, Washington State University; John Funkhouser, State University of New York, Stony Brook; P. W. Gast, Columbia University; W. R. Greenwood, MSC; Maurice Grolier, USGS; C. S. Gromme, USGS; G. H. Heiken, MSC; W. N. Hess, MSC; P. H. Johnson, BRN; Richard D. Johnson, Ames Research Center; Elbert A. King, Jr., MSC; N. Mancuso, MIT; J. D. Menzies, USDA; J. K. Mitchell, University of California at Berkeley; D. A. Morrison, MSC; R. Murphy, MIT; G. D. O'Kelley, ORNL; G. G. Schaber, USGS; O. A. Schaeffer, State University of New York, Stony Brook; D. Schleicher, USGS; A. H. Schmitt, MSC; Ernest Schonfeld, MSC; J. W. Schopf, University of California at Los Angeles; R. F. Scott, California Institute of Technology; E. M. Shoemaker, California Institute of Technology; B. R. Simoneit, University of California at Berkeley; D. H. Smith, University of California at Berkeley; R. L. Smith, USGS; R. L. Sutton, USGS; S. R. Taylor, Australian National University; F. C. Walls, University of California at Berkeley; Jeff Warner, MSC; Ray E. Wilcox, USGS; J. Zähringer, Max-Planck-Institute, Heidelberg, Germany.

The members of the Lunar Sample Analysis Planning Team are W. N. Hess (Chairman), Physicist, MSC; J. R. Arnold, Geochemist, University of California, San Diego; G. Eglinton, Geochemist, University of Bristol, England; C. Frondel, Mineralogist, Harvard University; P. W. Gast, Geochemist, Columbia University; E. M. Shoemaker, Geologist, California Institute of Technology; M. G. Simmons, Geophysicist, MIT; R. M. Walker, Geophysicist, Washington University; G. J. Wasserburg, Geochemist, California Institute of Technology; L. P. Zill, Biologist, Ames Research Center.

6. Passive Seismic Experiment

*Gary V. Latham, Maurice Ewing, Frank Press,
George Sutton, James Dorman, Nafi Toksoz,
Ralph Wiggins, Yosio Nakamura, John Derr, and Frederick Duennebier*

Purpose of the Passive Seismic Experiment

The Passive Seismic Experiment Package (PSEP), which was deployed on the lunar surface by the Apollo 11 astronauts, was the principal tool for determining the internal structure, physical state, tectonic activity, and composition of the Moon. Detailed investigation of the lunar structure must await the establishment and operation of a network of seismic stations; however, a single suitable well-recorded seismic event can provide information that is of fundamental importance and that could not be gained in any other way.

The PSEP flown on the Apollo 11 mission offered the first opportunity for demonstrating the feasibility of lunar seismic exploration. Seismometers are among the most delicate of scientific instruments. They are normally operated on piers in underground vaults at sites especially selected for environmental stability and a low level of background noise. The impossibility of attaining such favorable operating conditions on the lunar surface posed many design problems. A principal goal of this experiment was to demonstrate the feasibility of obtaining seismic records of motions of the lunar surface and to interpret them in terms of lunar structure. The problem to be resolved was to produce seismometers that could be installed with a minimum demand on an astronaut's time and energy, be adjusted and calibrated by remote control, be operated on a foundation with uncertain mechanical properties and with sensitivity characteristics that could be adjusted to conform with the background noise characteristics of the Moon, and that could remain operable through the large changes in temperature on the lunar surface.

The PSEP was successfully installed at Tranquility Base as part of the Apollo 11 mission and was operated for 21 days. It demonstrated that the goals of lunar seismic exploration could indeed be achieved.

Instrument Performance

The PSEP is shown schematically in figure 6–1. The system weighs 48 kg and uses solar cells to supply power. Thus, operation of the PSEP is confined to lunar day. The PSEP sensor unit contains three long-period (LP) seismometers with resonant periods of 15 sec, alined orthogonally to measure surface motion in both horizontal and vertical directions, and a single-axis, short-period (SP) seismometer sensitive to vertical motion at higher frequencies (resonant period of 1 sec).

Low-frequency, horizontal-component seismometers are extremely sensitive to tilt and must be leveled to high accuracy. In the Apollo seismic system, the two low-frequency, horizontal-component seismometers are leveled to within 2 seconds of arc from the local vertical direction by means of a two-axis, motor-driven gimbal (one motor for each seismometer). A third motor adjusts the low-frequency, vertical-component seismometer in the vertical direction. Motor operation is controlled from the Earth. These elements are shown schematically in figure 6–2. The low-frequency seismometers are mounted in crisscross fashion to achieve a minimum volume configuration. The instrument is controlled from the Earth by a set of 15 commands that control such functions as speed and direction of the leveling motors, instrument gain, and calibration.

Seismic data were obtained from the Apollo 11 seismometer system for 21 days. Initial activa-

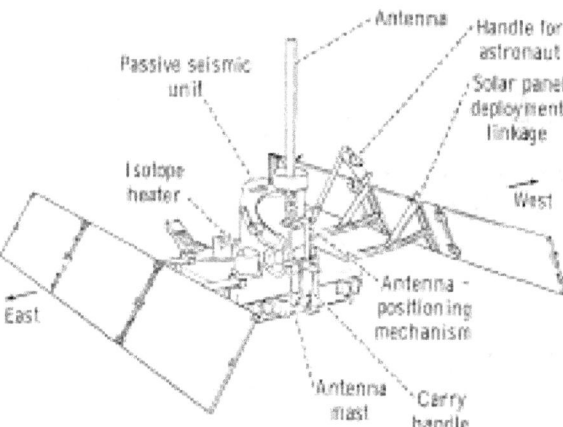

FIGURE 6-1. — Diagram of the fully erected PSEP.

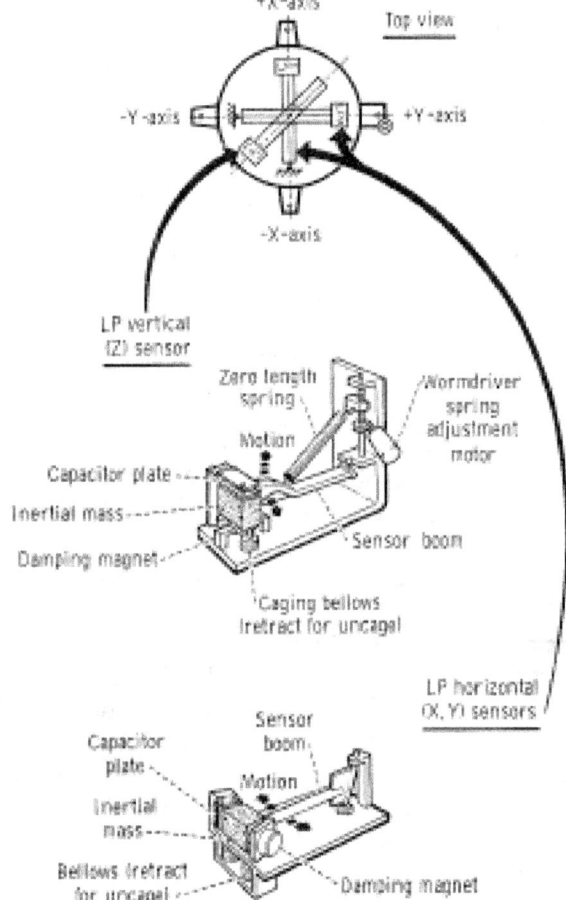

FIGURE 6-2. — Schematic diagrams of the elements of the low-frequency (LP) seismometers.

tion was on July 21, 1969, and final turnoff was on August 27, 1969. The instrument was emplaced south of the lunar module (LM), 16.8 m from the nearest point of the LM, as shown in figure 3–16. The PSEP is shown erected on the lunar surface in figure 6–3.

Termination of the experiment on August 27 resulted from failure of the system to receive and execute commands from Earth. Analysis of calibration pulses and signals received from various astronaut activities indicates that all four seismometers operated properly until the final 2 days of operation when the LP seismometers drifted out of their operating ranges and could not be recentered because of the absence of command capability. Instrument response curves, as derived from calibration pulses, are shown in figure 6–4.

Aside from the eventual failure of the system to respond to commands, the most significant deviation from nominal operating characteristics was that the actual maximum instrument temperature (190° F) exceeded the planned maximum (140° F) by 50° F. This resulted in occasional transient signals on the LP seismometer channels and several other minor effects, but did not significantly degrade instrument performance. The PSEP performance fully demonstrated the feasibility of operating seismic instruments in the lunar environment.

One problem that became evident in the early stages of the experiment was that the LM is a source of seismic signals of unexpectedly large amplitudes. Such seismic signals are assumed to be generated primarily by venting gases or circulating fluids, or both, within the LM. Inelastic deformation of the LM structure in response to thermoelastic stress may also be a significant source of seismic noise. Such noise signals make interpretation of the data obtained from the experiment more difficult because they must be recognized and separated from the signals of events of natural origin that are sought.

The problem of seismic noise from the LM will be reduced in later missions when the seismometers can be deployed at greater distances from the LM. Because of this problem, every effort must be made to achieve the greatest possible separation between the LM and the seismometers.

FIGURE 6-3.—Photograph of the PSEP immediately after its deployment on the lunar surface.

Description of Recorded Seismic Signals

LP Seismometers

Aside from occasional transient signals of instrument origin, the most conspicuous signals observed on the seismograms from the LP seismometers are a series of wave trains having the appearance of surface waves, that is, waves that would propagate along the surface of the Moon in contrast to body waves, which travel through the lunar interior. A total of 30 such signals can be clearly identified. No signals that might be classified as body waves are observed in relation to these wave trains. Most of the trains begin with SP (2 to 4 sec) oscillations that gradually increase to periods of 16 to 18 sec; that is, the

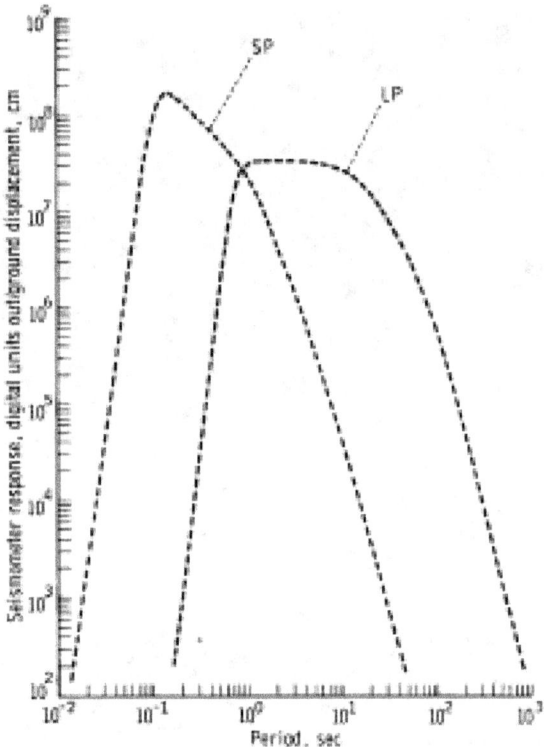

FIGURE 6-4. — Response curves for the SP and LP seismometers (for the highest gain setting). Response curves for the three LP seismometers are matched to within 5 percent. One digital unit is equal to 5 mV.

trains are dispersed. The first of these recorded wave trains is shown in figure 6-5. This wave train is primarily on the x-component. For several wave trains, the period range is between 20 and 50 sec. No corresponding signals are found on the LP vertical-component seismograms. Dispersion curves (wave period versus arrival time) for the first three wave trains are shown in figure 6-6.

During the second lunar day that the experiment was operational, several of these wave trains were observed to occur simultaneously with a series of pulses on the SP vertical seismometer. The pulses on the SP vertical seismometer are of uniform amplitude, and the time interval between pulses gradually increases. In the situations where wave trains and pulses on the SP vertical seismometer occur simultaneously, it is evident that the LP wave trains are simply the summation of the transient signals that correspond to the pulses. Amplitudes are of the order of 10 mμm peak-to-peak, so these are extremely small signals. While a natural origin for these signals cannot presently be excluded, the evidence strongly suggests that this entire set of signals is produced by instrumental effects that are, as yet, unexplained.

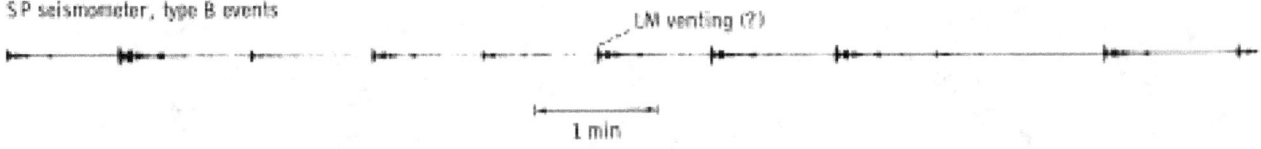

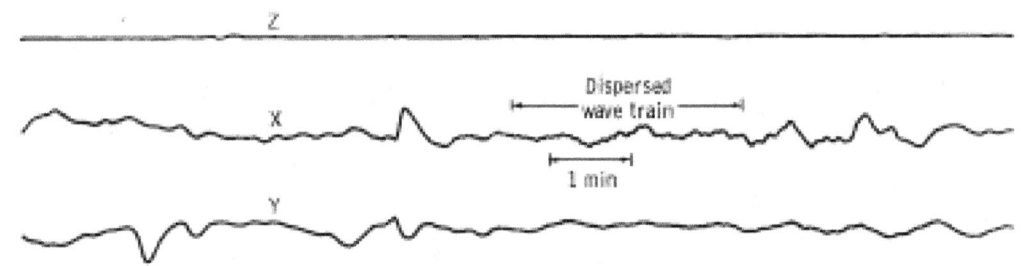

FIGURE 6-5. — Sample of a high-frequency seismic signal, perhaps generated by the LM, recorded by the SP seismometer and a sample of the background noise on the LP seismometer channels. This section of the seismic record includes one of the dispersed wave trains.

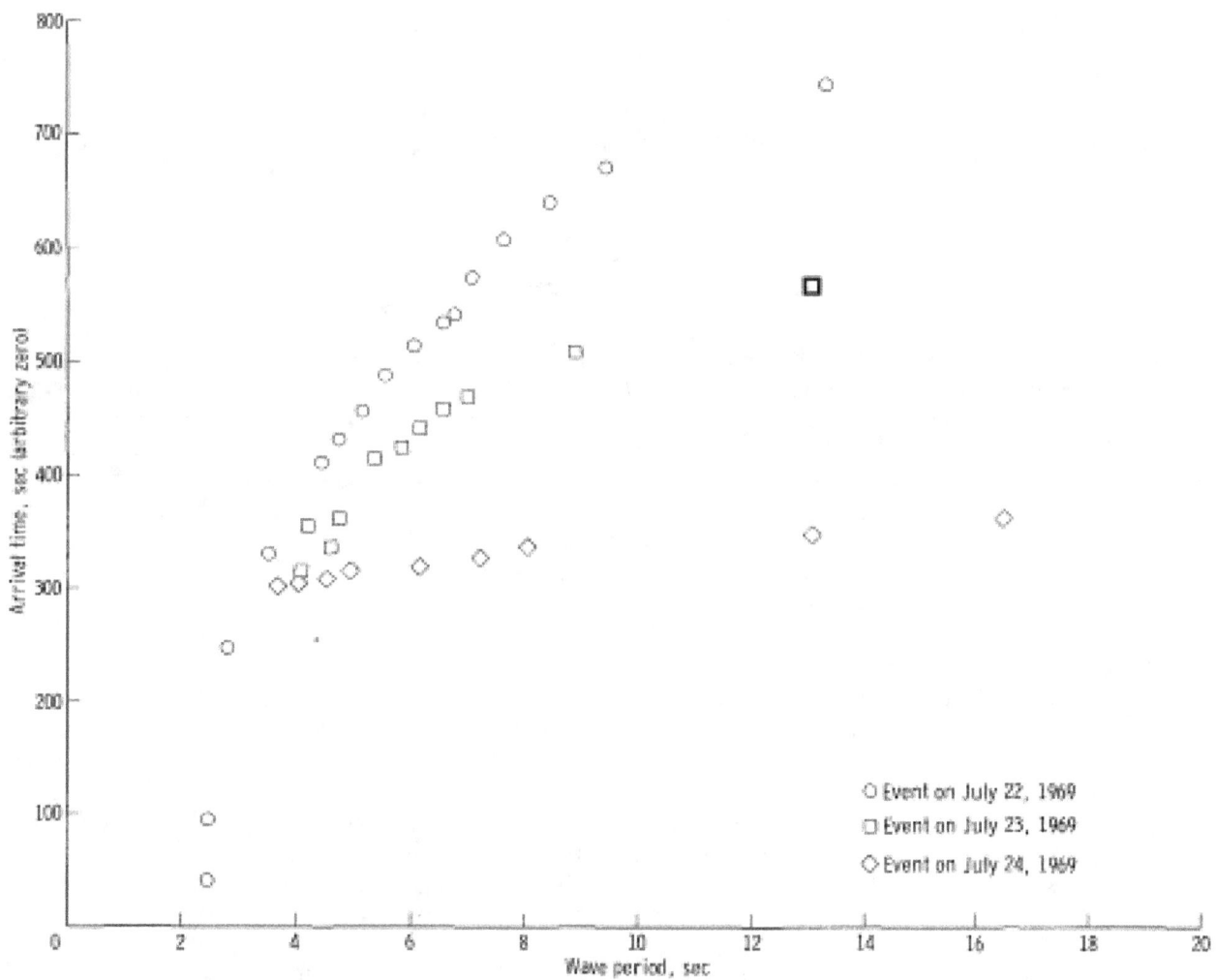

FIGURE 6-6.— Dispersion curves (wave train period versus arrival time) for three of the dispersed wave trains recorded by the SP seismometer.

The SP Seismometer

A great variety of high-frequency (2 to 18 Hz) signals is present on the seismograms from the SP seismometer. As a part of the preliminary investigations, an attempt has been made to sort these signals into descriptive categories based primarily upon the shapes of the signal envelopes and the spectral characteristics of each type. Only a few spectra are available at the present time. The signals are divided into the following categories:

(1) Signals produced by astronaut activities while the astronauts were on the lunar surface (type A)

(2) Signals with impulsive onset and relatively short duration (types I and X)

(3) Signals with emergent onset and relatively long duration (types B, M, T, and L)

Refinement in this categorization process can be expected as the investigation proceeds. Such refinement will be conducted with a view toward a better understanding of the source mechanisms involved.

Type A. Signals produced by astronaut activities were prominent on the SP seismometer from initial operation until LM ascent. Such signals were particularly strong when the astronauts were in physical contact with the LM. The signal produced when Astronaut Neil A. Armstrong ascended the ladder to reenter the LM is shown

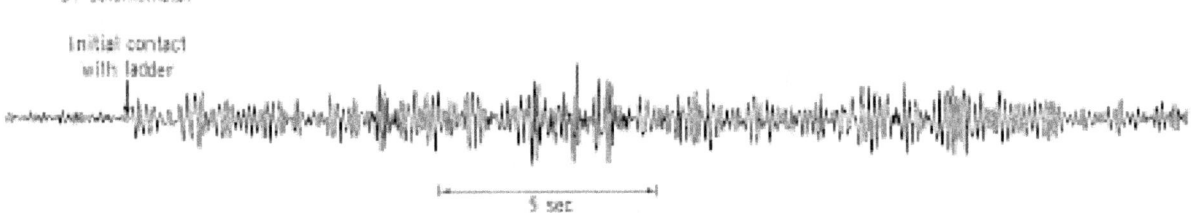

FIGURE 6-7. — Signal recorded when Astronaut Armstrong ascended the ladder of the LM.

in figure 6-7. The spectra of signals from various astronaut activities are shown in figure 6-8. The predominant frequency of all of the type A signals is between 7.2 and 7.3 Hz. This predominant frequency is assumed to correspond to a fundamental mode of vibration of the LM. Three items that were ejected from the LM struck the surface, and the spectra received also contain the 7.2-Hz peak (fig. 6-9). (These items were two portable life support systems, weighing 75 lb each, and the LM armrests, weighing a total of 12 lb.) However, it is important to note that the peaks near 11 and 13 Hz in figure 6-9 would be dominant if the spectra were corrected for instrument response. The spectral peak at 7.2 Hz, perhaps caused by LM resonance, was generated presumably because each ejected item struck the LM porch and ladder while falling to the surface.

In the spectra of signals recorded after the departure of the LM ascent stage, the spectral peak near 7.2 is shifted to 8.0 Hz. Resonances in the LM descent-stage structure, which was left on the lunar surface, would be expected to shift to higher frequencies when the mass of the ascent stage was removed.

The spectrum of one sample of signals from astronaut activity that was performed while the astronaut was not in contact with the LM is included in figure 6-8 for comparison with activities that were performed in physical contact with the LM. In the spectrum of sample (h) of figure 6-8, Astronaut Armstrong was collecting rock samples in the vicinity of the seismometers. The relative spectral content at higher frequencies (10 to 15 Hz) is greater for the sample (h) spectrum than for the other samples. Seismic signals in this frequency range can thus be attributed to astronaut footfalls on the lunar surface.

Type B. When the PSEP was turned on, type B signals with very large amplitudes were present on the SP seismometer record. The signals gradually decreased over a period of 8 days, until they disappeared completely on July 29, 1969. Maximum signal levels of approximately 200 mμm were recorded during the initial stages of activity. Type B signals are shown in figures 6-5 and 6-10. The sections of the record that are shown in figure 6-10 were recorded during the final stages of activity. After a puzzling sequence of changes in pattern, the signals became repetitive with nearly identical structure from train to train. The interval between events generally decreased to approximately 30 sec near their termination. Type B signals reappeared for a period of 13 hr during the second lunar day (13:00 hr on August 21, 1969, to 02:00 hr on

FIGURE 6-8. — Spectra of signals recorded by the SP seismometer during various astronaut activities: (a) Astronaut Aldrin ascending LM ladder, (b) Astronaut Armstrong ascending LM ladder, (c) reaction-control-thruster test firing (first quad), (d) reaction-control-thruster test firing (second quad), (e) impact of the second portable life support system, (f) impact of the first portable life support system, (g) impact of the LM armrests, and (h) Astronaut Armstrong collecting samples in the vicinity of PSEP. Spectra are not corrected for instrument response.

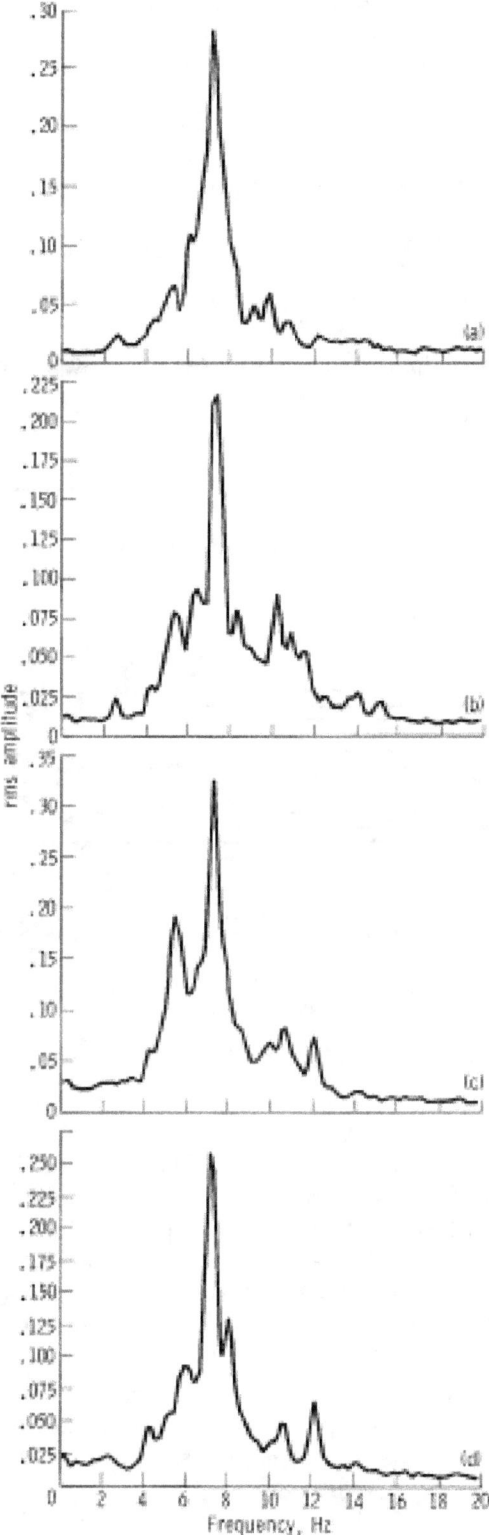
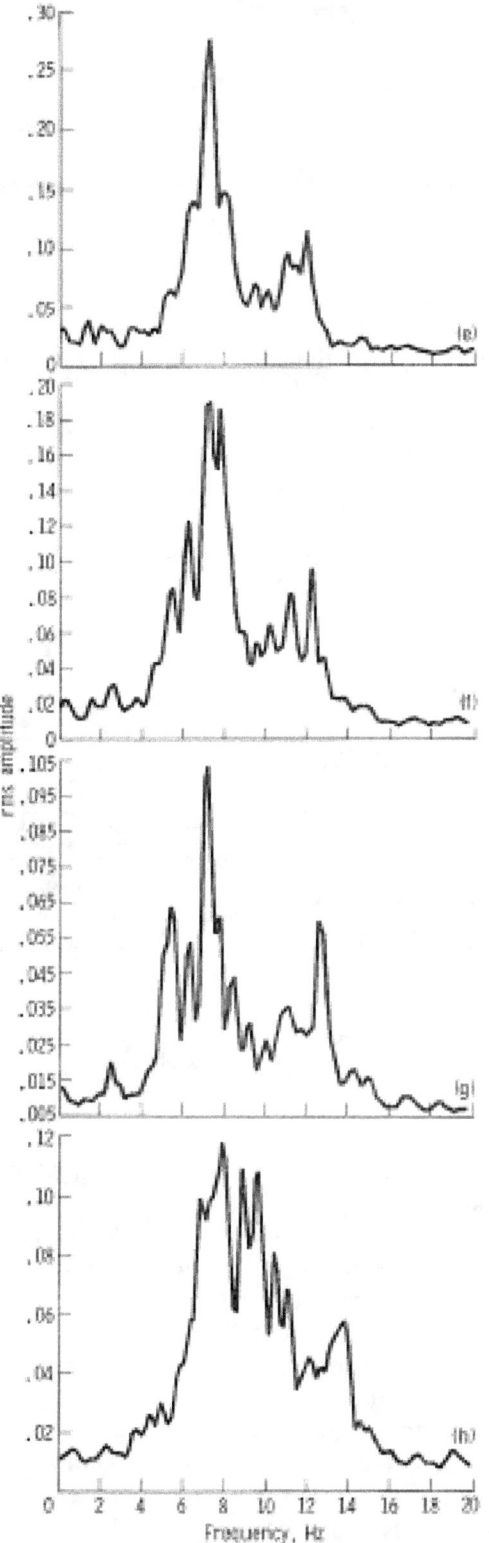

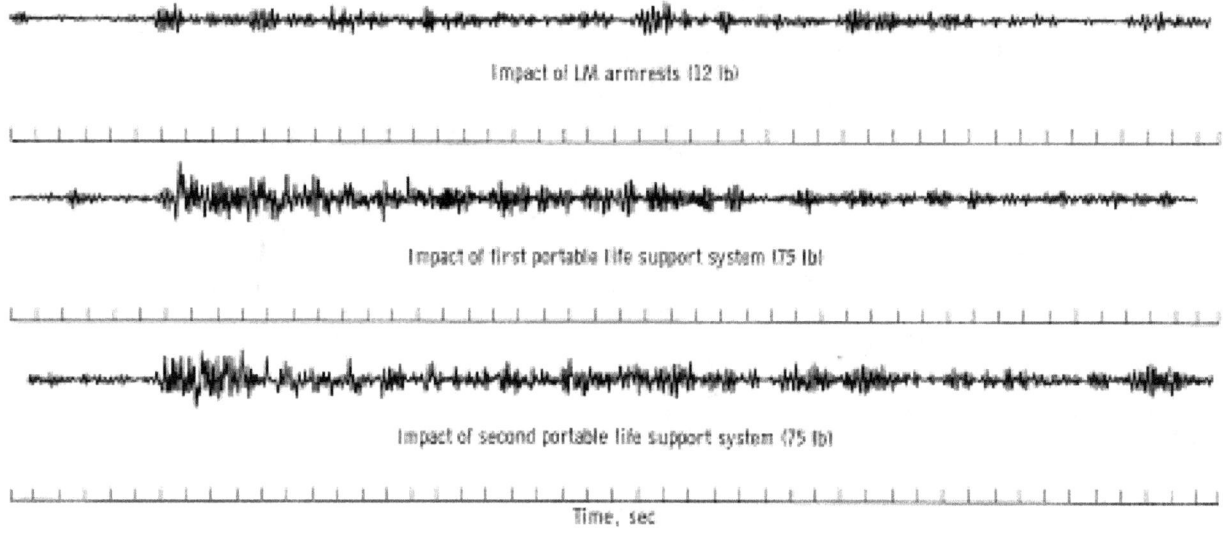

FIGURE 6-9.—Seismic signals recorded by the SP seismometer when various objects ejected from the LM struck the lunar surface.

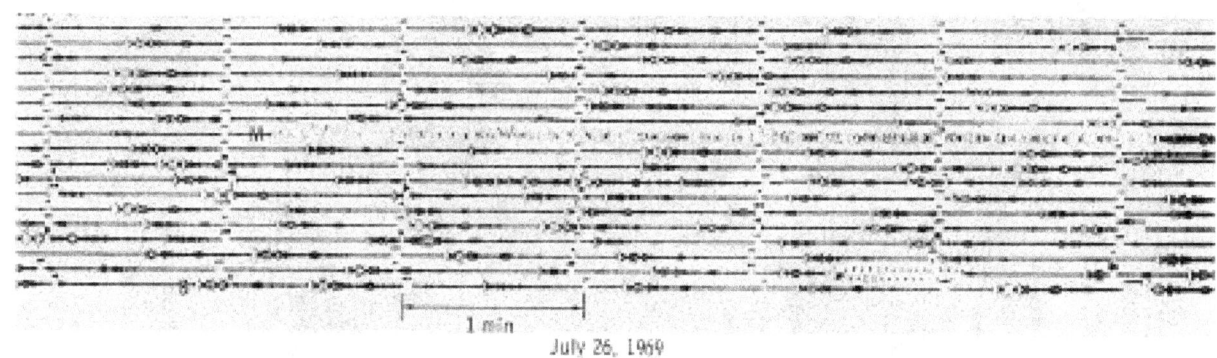

FIGURE 6-10.—Real-time record section from the SP seismometer showing type B and type M signals.

August 22, 1969) that the experiment was operational.

Spectra of three type B signals (and three type M signals) are shown in figure 6-11. The dominant spectral peak is 8.0 Hz. A subordinate peak at 6.5 Hz is also suggested. These type B events occurred after LM ascent. Before ascent, the predominant frequency was approximately 7.2 Hz. As mentioned previously, this shift in resonant frequency can be explained as the result of removing the mass of the ascent stage from the LM. The repetitive character of type B events, the apparent correlation of type B signals with LM resonant frequencies, and the eventual disappearance of type B signals have led to the tentative conclusion that the signals were produced within the LM structure, presumably by venting or circulating gases or liquids, or both.

Type M. In the 34-hr interval between 23:00 hr on July 25, 1969, and 09:00 hr on July 27, 1969, 14 seismic signals with unusually large amplitudes and long duration were recorded. These signals began impulsively and lasted up to 7 min for the largest trains. Low frequencies (one-tenth to one-fifteenth Hz) associated with the largest of these wave trains were also observed on the LP vertical-component seismometer (LPZ) records. No related signal was

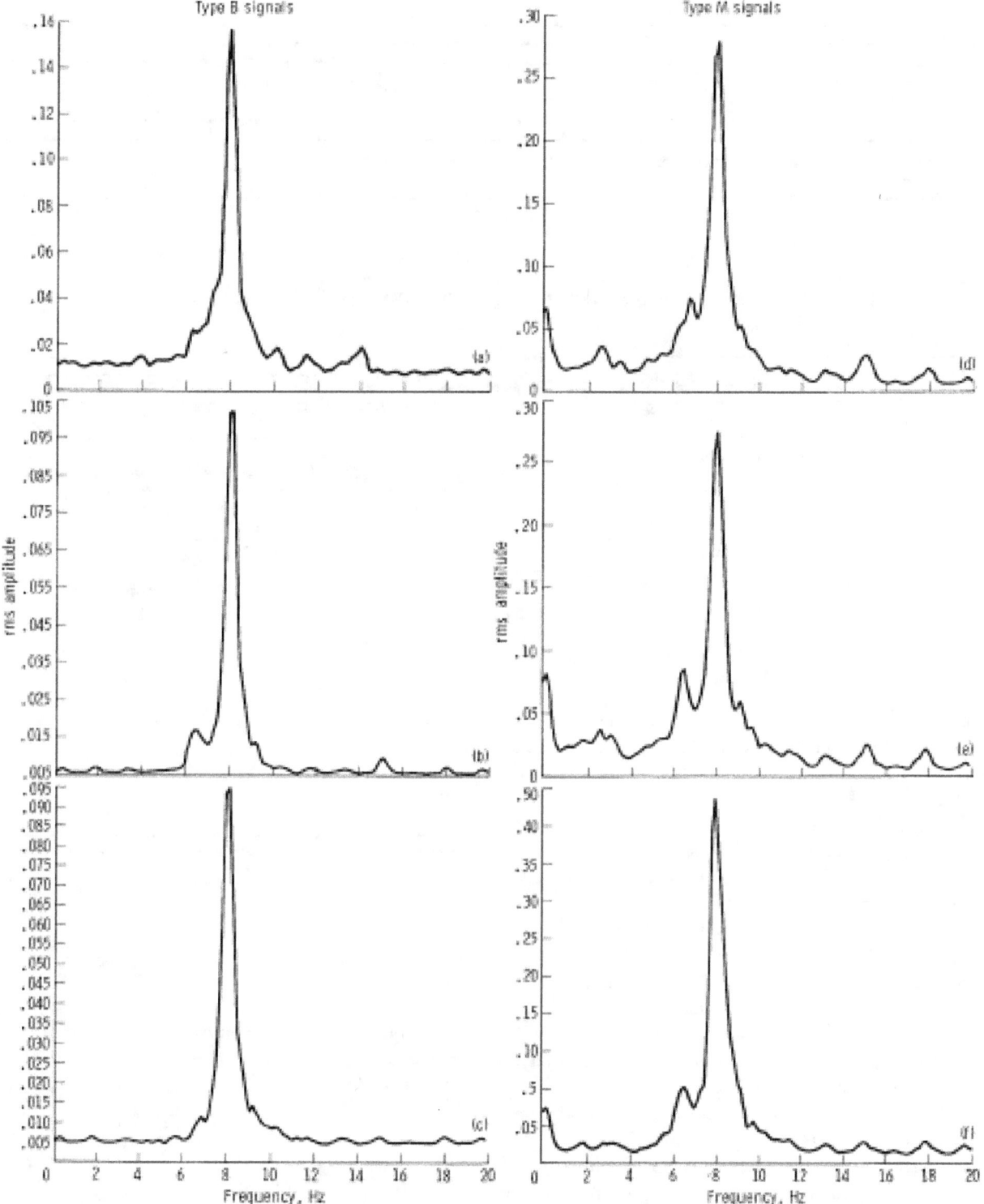

FIGURE 6-11.—Spectra of type B signals (samples (a), (b), and (c)) and type M signals (samples (d), (e), and (f)). Spectra are not corrected for instrument response.

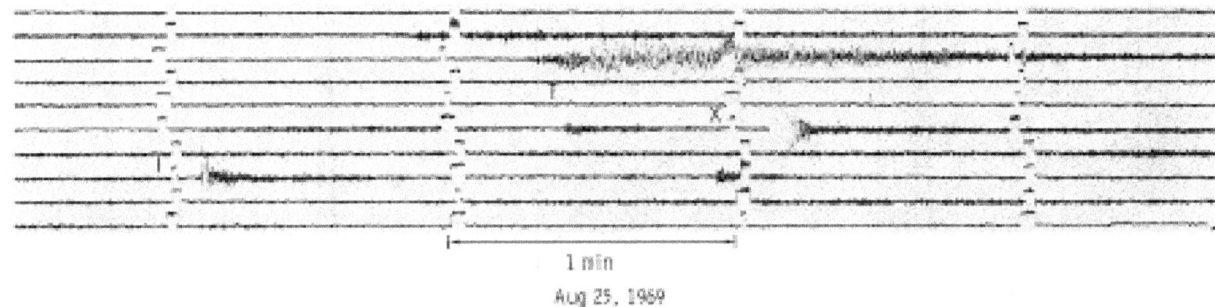

FIGURE 6-12. — Real-time record section of type T, I, and X signals from the SP seismometer.

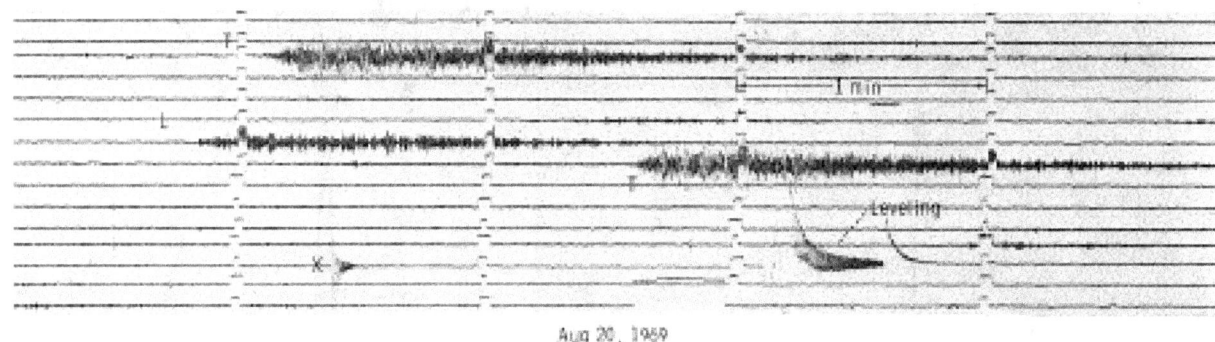

FIGURE 6-13. — Real-time record section of type T, L, and X signals from the SP seismometer.

observed on the horizontal components. The type M signals are believed to represent a separate class of events. A type M signal is shown in figure 6-10. Spectra for three type M events are shown in figure 6-11. These spectra are remarkably similar to the spectra of the type B events; hence, type M events may also have been generated within the LM.

Type T. Type T signals have emergent beginnings and reach maximum amplitudes within 10 sec after initial motion. The train gradually decreases in amplitude and has a total duration (above the 1-mm trace amplitude) of 1 to 4 min. Several samples of type T signals are shown in figures 6-12 and 6-13. The interval between signals is quite regular for long periods of time with very repetitive waveforms. In detail, the waveform changes with time in a complicated fashion. Type T events ceased completely on August 25, 1969.

Spectra for five type T signals are shown in figure 6-15. All type T event spectra show a broad peak near 8 Hz and a sharp peak between

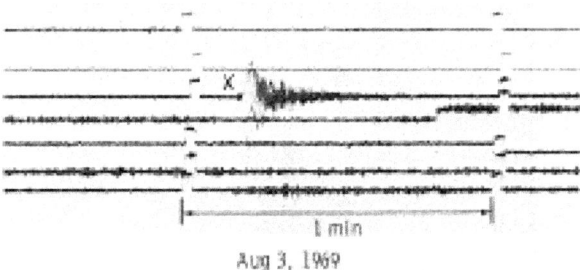

FIGURE 6-14. — Real-time record section of a type X signal from the SP seismometer.

12.8 and 14.7 Hz. A prominent peak near 5.7 Hz is contained in four of the spectra. When corrected for instrument response, the peak between 12.8 and 14.7 Hz would dominate all spectra.

Type I. Type I signals have impulsive beginnings and relatively short durations (less than 1 min). An example of a type I signal is shown in figure 6-12. No systematic pattern in the time of occurrence has been found for this type of signal. Spectra of seven type I signals are shown in figure 6-16. Samples (a), (b), and (c) in

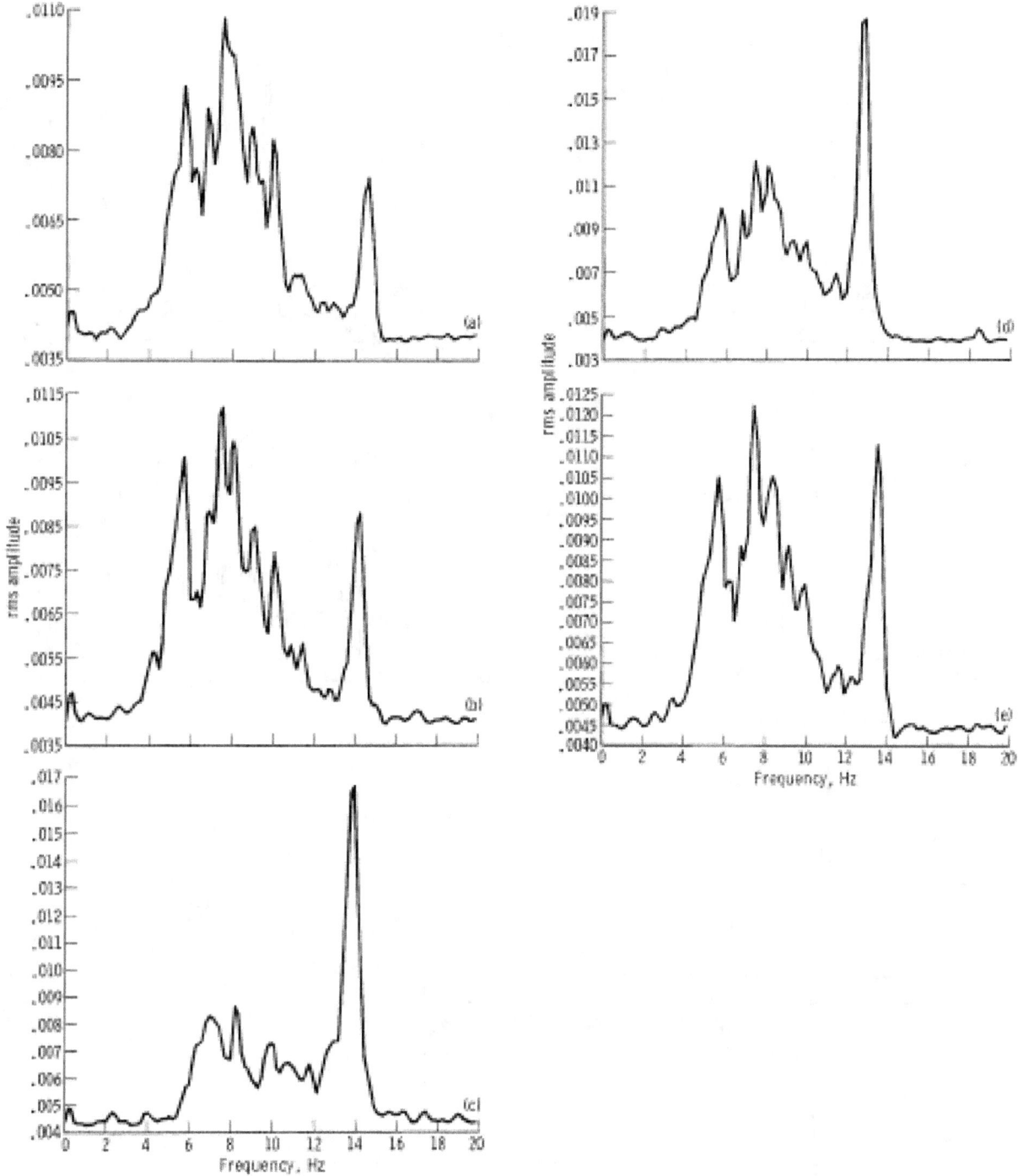

FIGURE 6-15.—Spectra of five type T signals. Spectra are not corrected for instrument response.

figure 6-16 are similar; all have a sharp peak near 8 Hz and a subordinate peak near 15 Hz. Samples (d), (e), and (f) in figure 6-16 are distinctly different from one another and from the first group. It is evident that several source mechanisms are involved in type I signals. Pos-

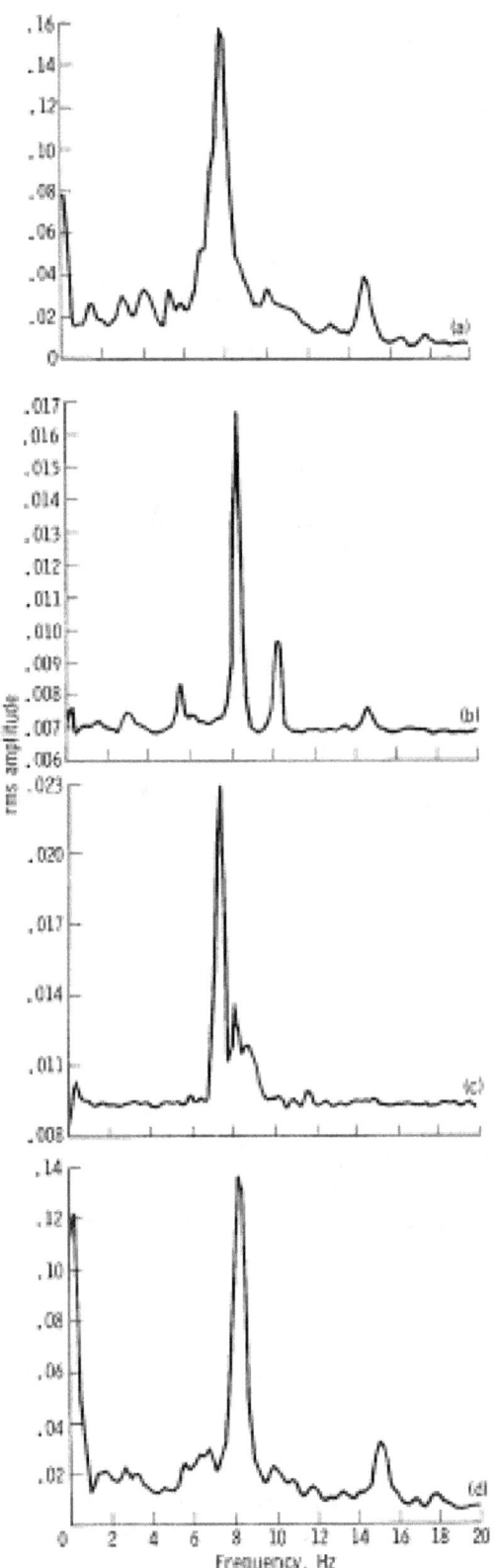
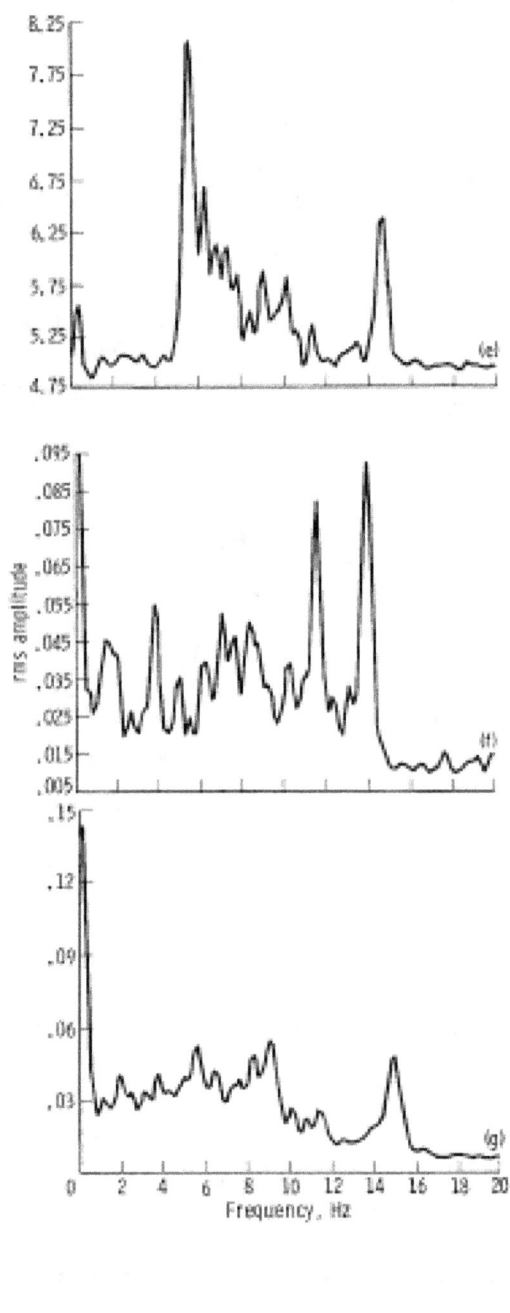

FIGURE 6-16. — Spectra of seven type I signals. Spectra are not corrected for instrument response.

sible sources for type I signals include (1) meteoroid impacts on the LM, on the lunar surface near the LM, and on the PSEP; (2) thermoelastic energy release within the LM and the PSEP; and (3) LM (and possibly PLSS) venting.

Type L. Type L signals have emergent beginnings and relatively long durations (1 to 6 min). The wave train builds up slowly and recedes slowly into the background, with maximum trace amplitude rarely exceeding 5 mm from peak to peak (approximately 3.3 mμm at 8 Hz). A type L signal spectrum is shown in figure 6–13. No regularity in the time of occurrence has been found for this type of signal. Of the various types of events, type L events contain the greatest variability in signal character and in time of occurrence. Spectra of three type L signals are shown in figure 6–17. As was noted for type I signals, type L signals show little similarity in spectral character for the few samples that have been computed. Thus, the presence of a variety of sources for type L events is inferred.

Type X. Type X signals have impulsive begin-

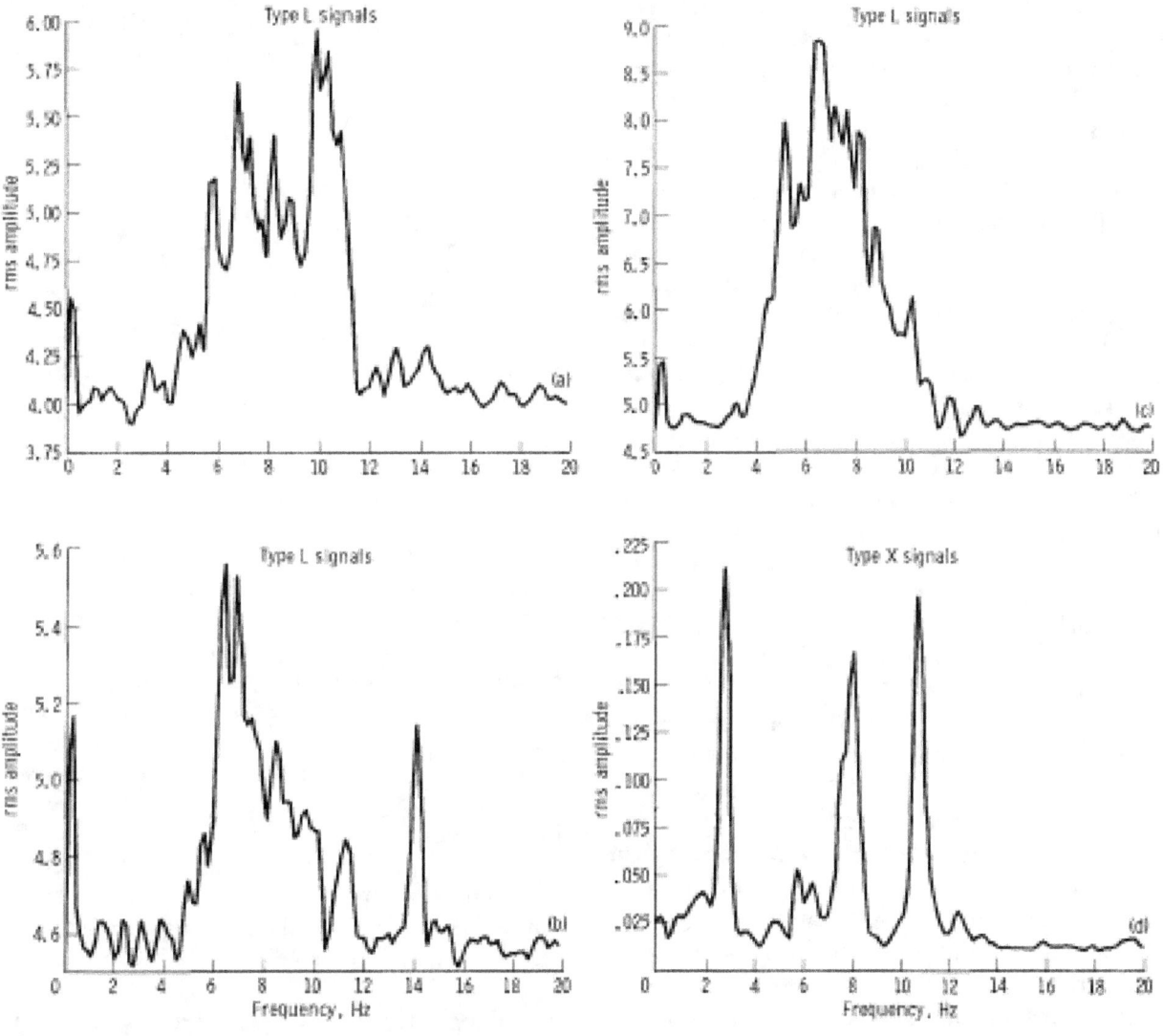

FIGURE 6-17. — Spectra of type L signals (samples (a), (b), (c)) and type X (sample (d)) signals. Spectra are not corrected for instrument response.

nings and relatively short durations (normally less than 10 sec). Maximum amplitudes occur at the beginning of the wave train and decay exponentially with a low-frequency (2 to 3 Hz) component that is usually evident in the tail of the train. Several examples of type X signals are shown in figures 6-12, 6-13, and 6-14. No regularity in the time interval between type X events has been found. The spectrum of a type X signal is shown in figure 6-17. Prominent peaks occur at 2.8, 8.0, and 10.8 Hz. When corrected for instrument response, the peaks at 2.8 Hz and at 10.8 Hz are dominant. The signal for which the spectrum (fig. 6-17) is shown occurred 38 hr after initial turnon, at the time when background signals of the B type were quite high; therefore, the 8.0-Hz peak may be the result of contamination associated with type B signals.

The spectral characteristic that clearly distinguishes type X signals from the other types of signals is the low-frequency peak (at 2.8 Hz). The most conspicuous structural resonance present within the PSEP in the 0- to 20-Hz band is associated with vertical oscillation of the solar-panel arrays. During development testing, the resonant frequency for this mode of oscillation was found to be 2.6 Hz. The coincidence of these frequencies and the clearly resonant character of the signal strongly support the hypothesis that solar-panel oscillation is the main contributor to type X signals. The most obvious sources for excitation of the solar-panel resonances are impacts of micrometeoroids on the PSEP and sudden dislocations produced within the PSEP structure by thermoelastic stress.

A histogram of activity for type X, I, L, and T signals is shown in figure 6-18. This figure represents the intensity of activity for each type of signal as measured on the SP seismometer. Logging of these different types of events was begun on the seismic records for July 28, 1969, when the level of periodic repetitive noises (type B signals) had subsided sufficiently so that type X, I, L, and T signals could be identified and compared with signals in later records. The vertical axis of each bar graph is a measure of the intensity of the activity. The width of each bar of the graph spans the time interval on each seismogram (usually 12 hr).

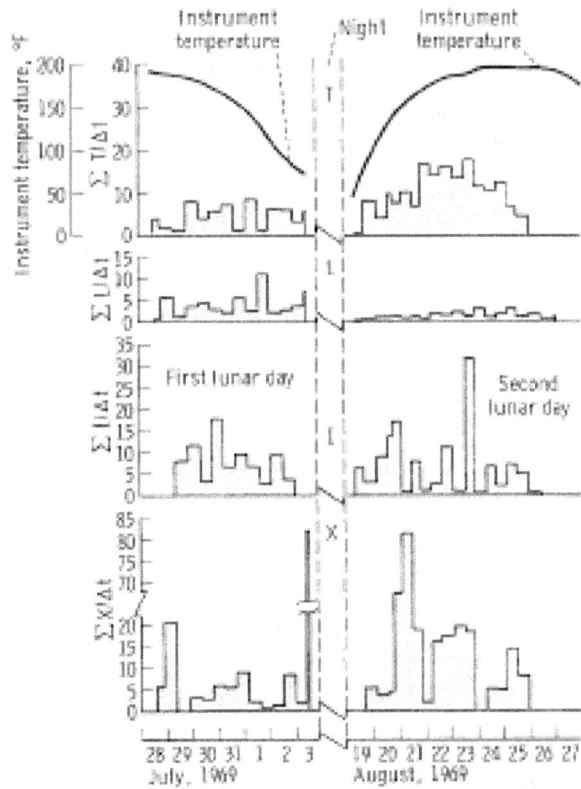

FIGURE 6-18. — A histogram of the intensity of activity for type X, I, T, and L signals.

The intensity of activity plotted in figure 6-18 is the summation of all peak-to-peak trace amplitudes greater than 1 mm on a particular seismogram, divided by the time interval t of the seismogram, in hours. Thus, the area of each bar of the graph is proportional to the total peak-to-peak trace amplitude, in millimeters, for all events of the particular type that occurred on the seismogram corresponding to that bar. The total area under a curve is, therefore, a measure of the cumulative intensity for the particular type of event. A curve showing central-station internal temperature is also given in figure 6-18.

As shown in figure 6-18, type X and T activities were more intense during the second lunar day of the experiment than during the first one, and type L activity declined slightly from the first lunar day to the second. Type I activity does not clearly change in a systematic way with time. There is a possibility that type X activity varies periodically, with the maxima of intensity that occurred at 2- to 3-day intervals during both the first and second lunar days the experiment was operational. A tendency exists for successive peaks of type X activity to decline during each lunar day. In figure 6–18, the highest peak of type X activity is reached soon after the beginning of the second lunar day. Type T activity is especially prominent during this second lunar day. It increased to a maximum and then disappeared in the interval between sunrise and local noon at Tranquility Base. Type T activity appears to have ended completely approximately 24 hr before the end of the recording period. Type X and I activities also appear to have ended at this time.

None of these types of activity (types T, X, and I) seems to be related to instrument temperature. Neither the absolute temperature nor the time derivative of the absolute temperature appears to govern the seismic activity. Known meteor streams do not appear to be related to any of the most conspicuous peaks of intensity; however, this possible relationship will be investigated further.

The possibility that some of the transient signals observed on the SP seismometer output are produced by meteoroids striking the PSEP can be examined. By applying the principle of conservation of momentum and using 10 kg as the mass involved in the motion of the instrument frame, 10 Hz as a representative signal frequency, 10 mμm as the maximum frame displacement (which is 20 percent of full scale for maximum gain at 10 Hz), and 30 km/sec as the velocity of the incoming meteoroid, a value of 2×10^{-7} is obtained for the meteoroid mass required to produce a transient signal of the average observed amplitude (10 mμm, peak-to-peak). The meteoroid mass required to produce the minimum detectable signal on the seismometer at 10 Hz is approximately 10^{-9} g. Using the meteoroid flux estimates of Hawkins (ref. 6–1), approximately five meteoroid impacts per day on the PSEP would produce seismic motions of approximately 10 mμm. This value is consistent with the hypothesis that type X events may be produced by direct meteoroid impacts on the PSEP.

At this point in the investigation, the following possible source mechanisms for the observed seismic signals are suggested:

(1) Venting gases from the LM and PLSS's and circulating fluids within the LM
(2) Thermoelastic stress relief within the LM and PSEP structures
(3) Meteoroid impacts on the LM, the PSEP, and the lunar surface
(4) Displacement of rock material along steep lunar slopes
(5) Moonquakes
(6) Instrumental effects

Feedback (Tidal) Outputs and Instrument Temperature

Signals corresponding to the slowly varying motion of each of the LP seismometers are transmitted on separate data channels. The data from these channels are referred to as feedback or tidal outputs. Instrument sensitivity at these outputs is sufficient to detect changes in gravity (on the vertical component) and the associated tilts of the lunar surface (on the horizontal components). In the PSEP, however, direct thermal effects on the instrument platform and on the spring of the LPZ seismometer were expected to greatly exceed changes associated with lunar tides. Internal instrument temperature was also transmitted as a separate data channel.

Variations in the feedback signals and instrument temperature are plotted in figure 6–19 for the total 21-day recording period. Step-function discontinuities in these signals are produced when the leveling and centering motors are in operation. These discontinuities have been removed from the output signal; hence, the total dynamic range of the seismometers is exceeded in the plot. Total excursion of the vertical component was equivalent to approximately 142 mgal. The total range in tilt of the horizontal-component seismometers was 370 μrad for the y-axis seismometer and 825 μrad for the x-axis seismometer. The expected peak-to-peak tidal variations are approximately 1 arcsec (4.9 μrad) of tilt and 1 mgal of gravity. As had been expected, these signals are much too small, relative to thermally induced signals, to be observed. As mentioned previously, the tilt variations are produced by a combination of actual tilting of the instrument platform in the lunar surface material, tilting associated with thermal distor-

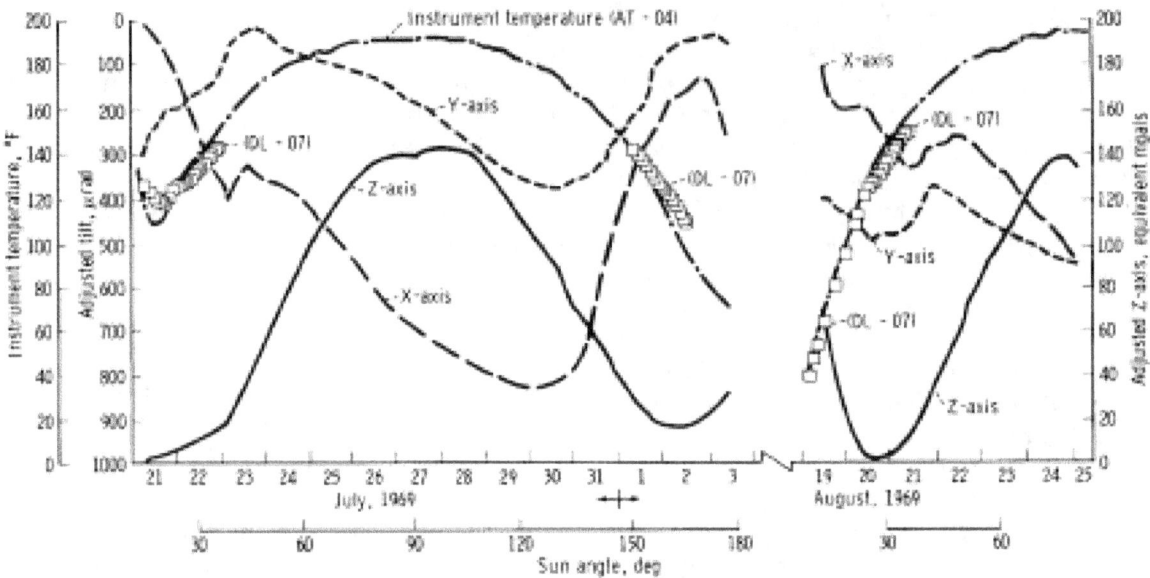

FIGURE 6-19. — Variations in feedback signals and instrument temperature over the 21-day recording period.

tions of the instrument platform, and direct thermal effects on the elements of the seismometers and on the seismometer leveling systems. Long-term variations in the LPZ seismometer feedback signal are produced primarily by direct thermal effects on the suspension spring.

Scientific Results

Perhaps the most important result of the PSEP is the demonstration that it is possible to explore another planet with the powerful tools of seismology. Despite the difficulty of interpreting the data in the presence of the background noises that resulted from the proximity of the experiment to the LM, sufficient information has been accumulated to show that greatly improved data can be obtained from the first Apollo Lunar Surface Experiments Package (ALSEP) by making simple changes in the operating procedure. Specific improvements in operating procedure that should be made include the following:

(1) The seismometer package must be moved farther from the LM to reduce interfering noise level caused by the LM.

(2) The sensor must be removed from the central station.

(3) Better thermal control must be achieved.

(4) Higher sensitivity must be sought.

All these improvements, except item (4), were planned for the Apollo 12 mission. Higher sensitivity of the seismic instruments must await the design changes that the advanced ALSEP will make possible.

Another important result of the PSEP is the discovery that the background noise level on the Moon is extremely low. Before these data were obtained, the level of lunar background noise was uncertain by several orders of magnitude. According to one popular hypothesis, the large diurnal thermal variations within the lunar surface material produced stresses that would lead to spallation and fracture and, consequently, to a high level of background noise. Another hypothesis proposed that meteoroid impacts were sufficiently numerous to create a significant continuous noise level. Still another hypothesis cited the absence of an atmosphere, oceans, and cultural activities to support the contention that the background noise level on the Moon should be much lower than the level on the Earth. The PSEP indicated that the background seismic-noise levels on the Moon are extremely low, as described in the following sections.

SP Background Seismic-Signal Level

A histogram of background seismic-signal level which was recorded by the SP seismometer for the first 9 lunar days of the experiment is shown in figure 6-20. The high-amplitude signal that occurs immediately after turnon is produced partly by astronaut activities and partly by signals that are tentatively attributed to the LM. These levels decreased steadily over a period of 8 days. After 8 days, occasional activity, as discussed in the section entitled "Description of Recorded Seismic Signal," was observed. Thus, the continuous background seismic signal near 1 Hz is less than 0.3 mμm, which corresponds to system noise. Maximum signal levels of 1.2 μm at frequencies of 7 to 8 Hz were observed during the period when the astronauts were on the surface.

LP Vertical-Component Seismometer

Except for the occasional occurrence of transient signals of instrumental origin, the background seismic-signal level on the LPZ seismometer is below system noise, that is, below 0.3 mμm over the period range from 1 to 10 sec. This level is between 100 and 10 000 times less than average background levels observed on Earth in the normal period range for microseisms (6 to 8 sec).

LP Horizontal-Component Seismometers

Continuous seismic background signals of extremely small amplitudes (10 to 30 mμm, peak-to-peak) were observed on the records from the two LP horizontal-component seismometers. The amplitude of these signals decreased considerably for a 2- to 3-day interval near lunar noon, when the rate of change of external temperature with time is at a minimum. The signals are very low frequency with periods of from 20 sec to 2 min. It is assumed that these signals correspond to the tilting of the instruments caused by a combination of thermal distortions of the metal pallet, which serves as the base of the seismometer, and a rocking motion of the pallet in the lunar surface material. The rocking motion could also be produced by thermal effects. However, the horizontal component of true lunar seismic-signal background level at shorter periods (less than 10 sec) also appears to be less than 0.3 mμm. The character of the background signals on the LP seismometers is shown in figure 6-5.

Of the many signals that have been recorded, at least some were produced by the LM. Many signals may be of natural origin; that is, they may have been generated by moonquakes, impacts, or movement of surface rocks. However, none of the observed signals has the pattern normally observed in recordings of seismic activity on Earth (with the possible exception of signals produced by local volcanic events and by landslides). Distinct phases corresponding to the various types of waves are not apparent in any of the recorded seismic signals, and most wave trains are of long duration. The high sensitivity at which the PSEP instruments were operated would have resulted in detection of many distinctive seismic events if the Moon were as seismically active as the Earth and if lunar rocks transmitted seismic waves as effectively as do Earth rocks. The fact that a large number of unmistakable seismic events were not observed during 21 days of operation is a major scientific result. To describe this observation, it is convenient to introduce the term "seismic receptivity."

Seismic receptivity is a measure of the overall effectiveness with which seismic waves are generated, transmitted, and detected at any given point for any given body. The seismic receptivity is the product of involving (1) the seismicity of the body, that is, the rate of seismic energy release, (2) the transmissibility of the medium through which the waves propagate, and (3) the level of background seismic noise at the sensor location.

The background seismic noise of the Moon (for the sensor at Tranquility Base) is low. This low background level results in high seismic receptivity for the sensor. The problem to be attacked by analysis of the data already obtained and data expected from future missions is to determine the extent to which each of two factors, seismicity and transmissibility, is responsible for the small number of distinctive seismic events that have been recorded.

Among the possible explanations of low seis-

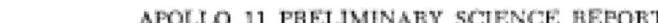

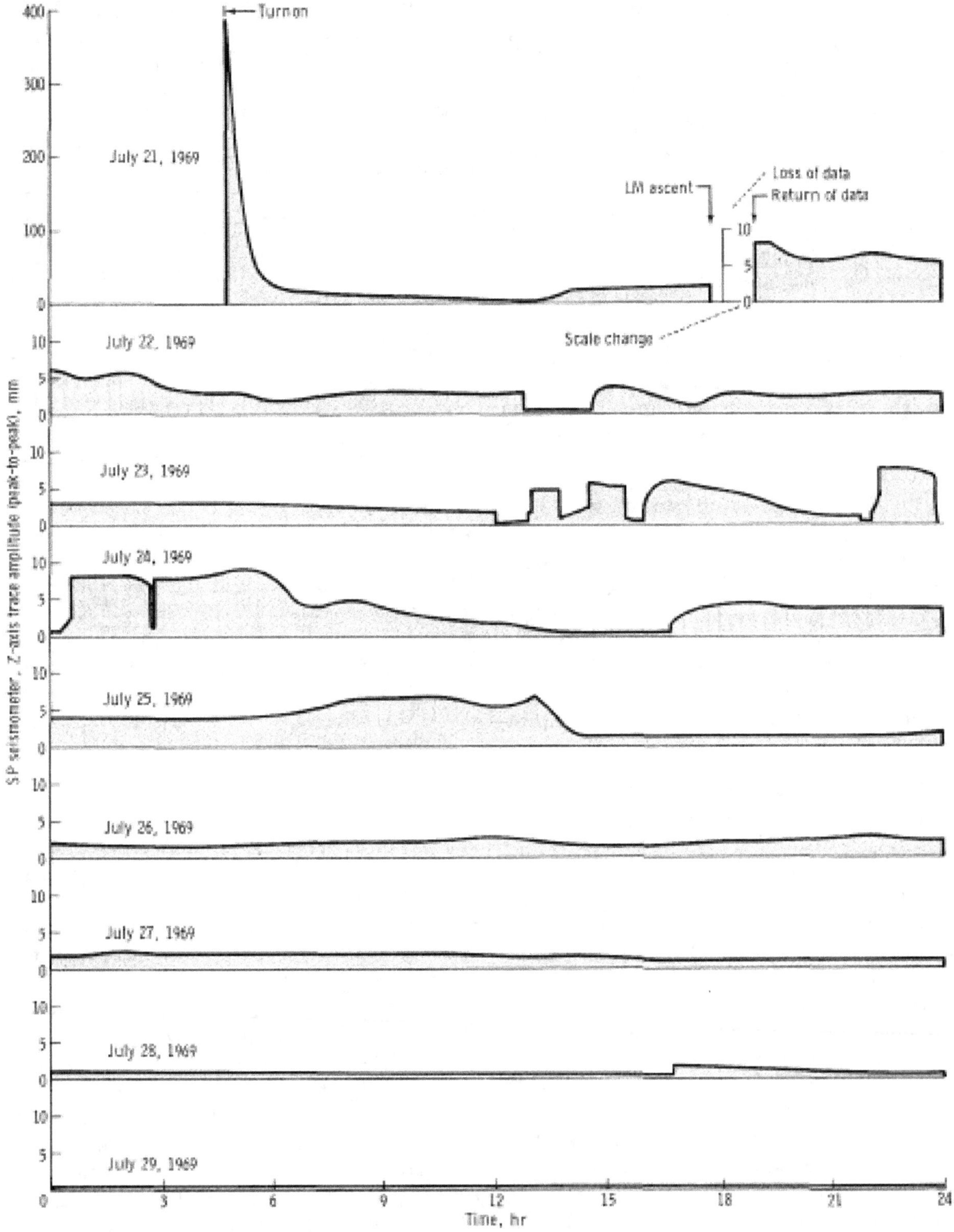

FIGURE 6-20. — Histogram of signal level from the SP seismometer. A 1-mm trace amplitude equals 1.9 mμm for the predominant signal component (8 Hz).

mic receptivity are the heterogeneity of lunar material, which would scatter the seismic waves; a low-Q (where Q is the transmissibility quality factor) interior, which would absorb seismic waves; and an inability of the lunar material to store high stress. Low seismicity, if confirmed by future missions, would imply the absence of tectonic processes within the Moon such as those associated with major crustal movements on Earth.

One hypothesis that could explain the observations is that the body of the Moon is very heterogeneous, at least in its outer regions. The near-surface material is certainly heterogeneous in character, as evidenced by the great numbers of visible craters and surface fractures. This heterogeneity may extend deep into the body of the Moon. The scattering of seismic waves that would occur in propagation through a highly heterogeneous material would tend to increase the duration of the observed seismic wave and to suppress the appearance of distinct phases within the wave train. The PSEP station was in one of the mare regions, which are thought to be great lakes of solidified lava. If this hypothesis is true, the maria would be expected to be among the most homogeneous regions of the Moon. However, since the original formation of the maria, meteoroid impacts may have converted these structures to rubble. If the age of Mare Tranquillitatis is as great as first results indicate (discussed in chapter 5), the process of fracturing by impact may have proceeded to a very great extent. Much of the visible evidence for such fracturing may be hidden by the overlying blanket of fragmental material. However, it should be noted that nothing in the present observations precludes the possibility of a high-temperature (low-Q) lunar interior. This would also contribute to low seismic receptivity.

With one lunar seismic station, one should not expect to do effective seismology on the Moon, unless the Moon were similar to the Earth, with an abundance of natural seismic sources, and unless the signals were similar in character to those observed on Earth, so that one could draw by analogy on Earth experience. Neither of these two criteria appears to have been met. Seismic experiments, nevertheless, will lead eventually to answers to questions concerning the structure and dynamics of the Moon. Results will come more slowly than had been hoped, and there will be greater dependence upon the establishment of a network of stations and use of artificial sources.

Meteoroid impacts are a major factor in shaping the lunar surface. Determination of the size and frequency distribution of meteoroid impacts is necessary to estimate quantitatively the rates of crater formation and erosion. A store of information, begun with the PSEP data, will be uniquely suited to the study of this problem.

Two important implications that will affect future seismic experiments emerge from the preliminary analysis of the PSEP data. The effectiveness of the use of moonquakes for investigation of deep lunar structure is low, despite the high sensitivity of the seismometers used. However, the low background noise observed demonstrates that greater sensitivity can be used. The effectiveness of the present instrumentation will be greatly enhanced when a planned network of three ALSEP stations is operable. Even if the seismicity of the Moon is considerably less than that of the Earth, this network can reasonably be expected to record several significant lunar seismic events. Data from even a single well-recorded lunar seismic event could be sufficient to determine major structural features of the Moon. Also, even if the Moon has an extremely low seismicity, probing of the deep lunar interior by use of artificial sources, such as the impacts of a Saturn IVB stage and the LM ascent stage, is a strong possibility.

The low seismic-noise level measured by the PSEP indicates that the major modification needed in the advanced ALSEP is increased sensitivity of the seismometer system. Increased sensitivity may compensate for the apparent low seismicity of the Moon to the extent that moonquakes can be as effective for investigation of deep lunar structure as earthquakes are for study of the interior of the Earth.

Reference

6-1. Hawkins, Gerald S.: The Meteor Population, Research Report No. 3. NASA CR-51365, August 1963.

7. Laser Ranging Retroreflector

C. O. Alley, P. L. Bender, R. F. Chang, D. G. Currie, R. H. Dicke, J. E. Faller, W. M. Kaula, G. J. F. MacDonald, J. D. Mulholland, H. H. Plotkin, S. K. Poultney, D. T. Wilkinson, Irvin Winer, Walter Carrion, Tom Johnson, Paul Spadin, Lloyd Robinson, E. Joseph Wampler, Donald Wieber, E. Silverberg, C. Steggerda, J. Mullendore, J. Rayner, W. Williams, Brian Warner, Harvey Richardson, and B. Bopp

Concept of the Experiment

The compact array of high-precision optical retroreflectors (cube corners) deployed on the Moon is intended to serve as a reference point in measuring precise ranges between the array and points on the Earth by using the technique of short-pulse laser ranging. The atmospheric fluctuations in the index of refraction prevent a laser beam from being smaller than approximately 1 mile in diameter at the Moon. The curvature of the lunar surface results in one side of the short pulse being reflected before the other side, producing a reflected pulse measured in microseconds, even if the incident pulse is measured in nanoseconds. The retroreflector array eliminates this spreading because of the small size of the array. (The maximum spreading of a pulse because of optical libration tipping of the array will be approximately ±0.125 nsec.) In addition, the retroreflective property causes a much larger amount of light to be directed back to the telescope from the array than is reflected from the entire surface area illuminated by the laser beam.

The basic uncertainty in measuring the approximately 2.5-sec round-trip travel time is associated with the performance of photomultipliers at the single photoelectron level. This uncertainty is estimated to be approximately 1 nsec. When the entire system is calibrated and the effects of the atmospheric delay are calculated from local temperature, pressure, and humidity measurements and subtracted from the travel time, where the uncertainty in this correction is estimated to be less than 0.5 nsec, an overall uncertainty of ±15 cm in one-way range seems achievable.

With the ±15-cm uncertainty, monitoring the changes in point-to-point distances from Earth to the lunar reflector (by daily observations for many years) will produce new information on the dynamics of the Earth-Moon system. The present uncertainty of three parts in 10^7 in the knowledge of the velocity of light will not affect the scientific aims of the experiment, since it is the practice to measure astronomical distances in light travel time. Primary scientific objectives include the study of gravitation and relativity (secular variation in the gravitational constant), the physics of the Earth (fluctuation in rotation rate, motion of the pole, large-scale crustal motions), and the physics of the Moon (physical librations, center-of-mass motion, size and shape). Some of these objectives are discussed in references 7–1 to 7–4. Estimates made by P. L. Bender of improvements expected in some of these categories are shown in tables 7–I to 7–III.

Properties of the Laser Ranging Retroreflector

Although the Laser Ranging Retroreflector (LRRR) is simple in concept, the detailed design of a device that would satisfy the stated scientific aims has received much attention. The primary design problem has been to avoid systematic gaps in ranging data expected to result from the extreme variation in thermal conditions on the Moon (from full Sun illumination to lunar night). A preliminary design based on discussions among various members of the investigator group and optical engineers was put

TABLE 7-I. Lunar orbital data parameters

Quantity	Present accuracy (approximate)	1.5-m range uncertainty		0.15-m range uncertainty[a]	
		Accuracy	Time, yr	Accuracy	Time, yr
Mean distance	500 m	250 m	1	75 m	0.5
				25 m	1
Eccentricity	1×10^{-7}	4×10^{-8}	1	1.5×10^{-8}	.5
				4×10^{-9}	1
Angular position of Moon with respect to perigee	2×10^{-6}	4×10^{-7}	1	1.5×10^{-7}	.5
Angular position of Moon with respect to Sun	5×10^{-7}	4×10^{-7}	1	1.5×10^{-7}	.5
				4×10^{-8}	1
Time necessary to check predictions of Brans-Dicke scalar-tensor gravitational theory, yr	---	25	25	8	8

[a] Three observing stations are assumed for periods longer than 0.5 yr.

TABLE 7-II. Lunar libration and relation of Laser Ranging Retroreflector (LRRR) to center of mass

Quantity	Present accuracy (approximate)	1.5-m range uncertainty		0.15-m range uncertainty[a]	
		Accuracy	Time, yr	Accuracy	Time, yr
Libration parameters:					
$\beta = (C-A)/B$	1×10^{-5}	3×10^{-7}	4	3×10^{-7}	0.5
				3×10^{-8}	4
$\gamma = (B-A)/C$	5×10^{-5}	2×10^{-6}	1.5	1.5×10^{-6}	.5
Coordinates of LRRR with respect to center of mass:				2×10^{-7}	1.5
X_1	500 m	250 m	1	75 m	.5
				25 m	1
X_2	200 m	70 m	1	40 m	.5
				7 m	1
X_3	200 m	50 m	3	50 m	.5
				5 m	3

[a] Three observing stations are assumed for periods longer than 0.5 yr.

TABLE 7-III. Geophysical data determinable from LRRR

Quantity	Present accuracy (estimated)	1.5-m range uncertainty	0.15-m range uncertainty
Rotation period of Earth, sec	5×10^{-3}	10×10^{-4}	1×10^{-4}
Distance of station from axis of rotation, m	10	3	0.3
Distance of station from equatorial plane,[a] m	20	6 to 20[b]	0.6 to 2[b]
Motion of the pole,[a] m	1 to 2	1.5	0.15
East-west continental drift rate observable in 5 yr,[a] cm/yr	30 to 60	30	3
Time for observing predicted 10-cm/yr drift of Hawaii toward Japan,[a] yr	15 to 30	15	1.5

[a] Three or more observing stations are required.
[b] Depending upon the latitude of the station.

forth by J. E. Faller in a proposal to NASA (ref. 7–5).

The first financial support provided by NASA was used to test the proposed design, which consisted of small solid-corner reflectors made of homogeneous fused silica. The test and evaluation facilities at the NASA Goddard Space Flight Center were used to simulate the lunar environment. These tests verified that a metal coating could not be used on the reflectors and showed that the failure of total internal reflection at off-axis angles posed serious problems in mounting the reflectors to maintain small temperature gradients during larger off-axis Sun angles (ref. 7-6).

The tests verified the predicted performance during lunar night and during direct Sun illumination within the total internal reflection region of angles.

The efforts of thermal and mechanical design engineers in close association with members of the investigator group solved the problems and led to the design shown in figure 7-1. The corner reflectors are lightly mounted on tapered tabs between Teflon rings, and are recessed by one-half their diameter into cylindrical cavities in a solid aluminum block. The predicted optical performance, based on thermal analyses under changing lunar conditions, is shown in figure 7-2.

The need for careful pointing of the LRRR toward the center of the Earth libration pattern is shown by figure 7-3, which displays the off-axis response of the recessed corner reflectors. (The curve is the result of averaging over azimuthal orientation and polarization dependence.) The motion of the Earth in the lunar sky because of the optical librations of the Moon is shown in figure 7-4 for the period July to October 1969.

The alinement in the east-west direction

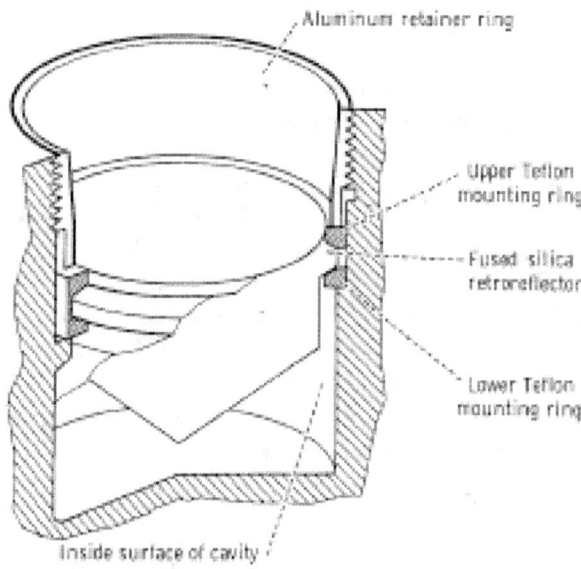

FIGURE 7-1. — Corner reflector mounting.

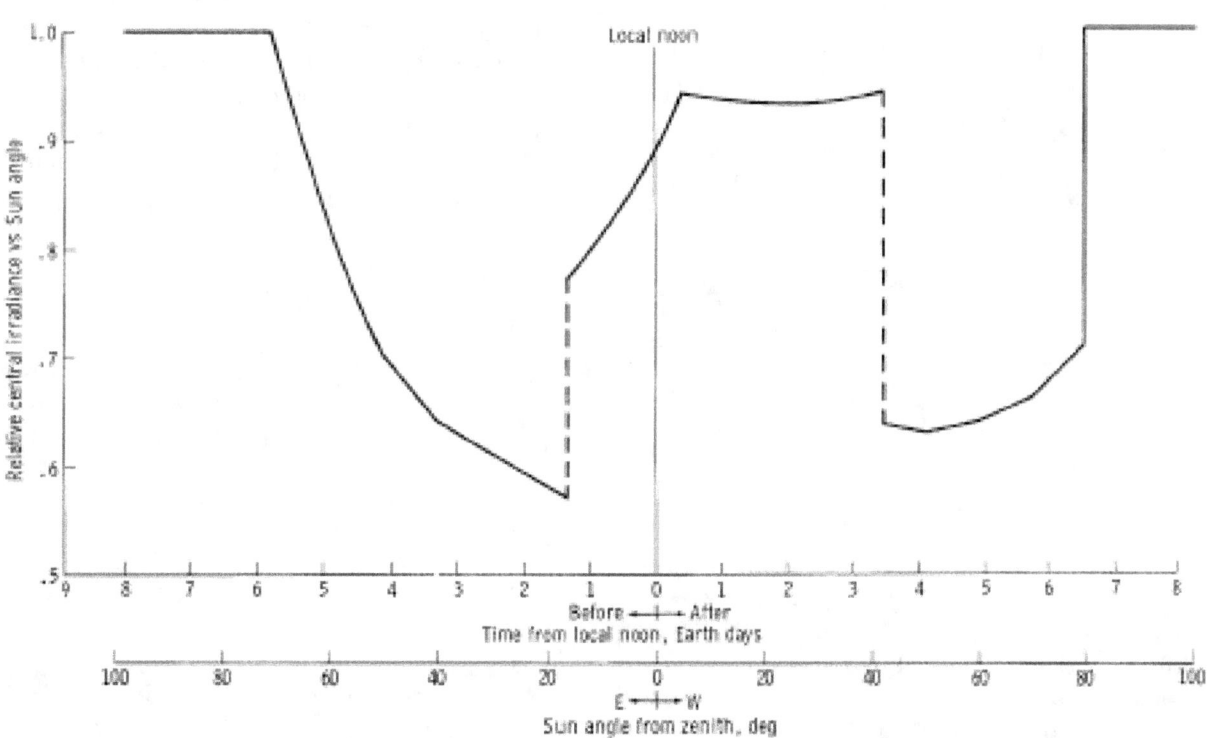

FIGURE 7-2. — Predicted optical performance as a function of Sun angle, from thermal analysis.

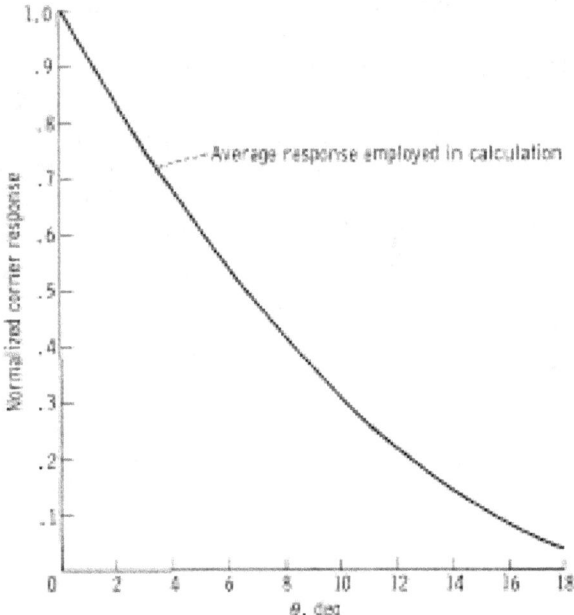

FIGURE 7-3. — Average off-axis performance of recessed circular corner reflectors.

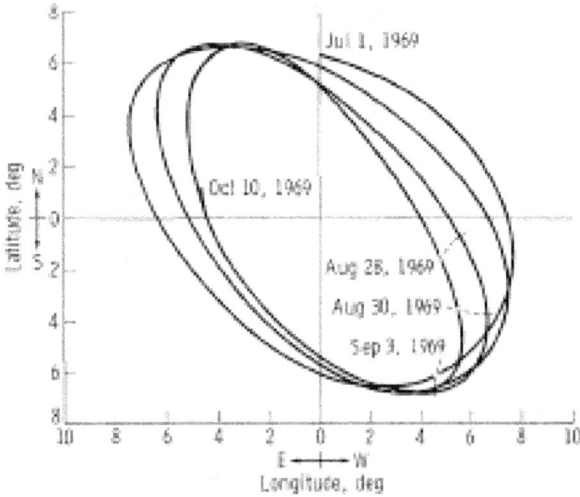

FIGURE 7-4. — Optical librations for the period July to October 1969.

achieved by Astronaut Neil A. Armstrong was within the width of a division on the compass mark. The leveling was within 0.5°, with the bubble oriented toward the southwest. When combined with the worst possible mechanical tolerances of construction, the east-west alinement is ±0.7°, and the overall pointing is within 1° of the center of the libration pattern.

This excellent pointing means that the return will not fall to extremely low values during the extremes of the libration cycle. With the previous information included, the relative expected response from July 20 to September 9, 1969, is shown in figure 7-5. Figure 7-5 is an upper bound for the performance expected.

The flight hardware is shown in figure 7-6. The gnomon, the alinement marks, and the bubble level are clearly shown. In figure 7-7, the LRRR is shown deployed on the Moon.

In tests, each corner reflector which was selected for the flight array exhibited an on-axis diffraction performance greater than 90 percent of that possible for a corner reflector having no geometrical or homogeneity defects. The on-axis, nondistorted performance is conveniently characterized by a differential scattering cross section.

$$\left. \frac{d\sigma}{d\Omega} \right|_{180°} = 5 \times 10^{11} \text{ cm}^2/\text{steradian}$$

which yields the number by which the photon flux density (photons per square centimeter) incident on the reflector array must be multiplied to give the number of photons per steradian intercepted by the telescope receiver. The cross section includes the effect of velocity aberration and is evaluated for a wavelength of 6943 Å.

Observations of returns on August 1 and 3, 1969 (immediately before lunar sunset), and on

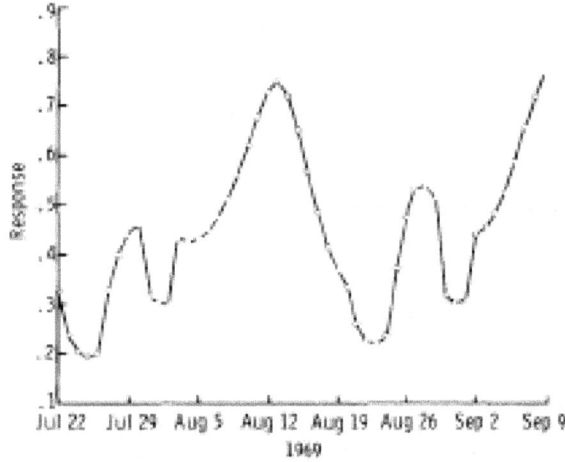

FIGURE 7-5. — Upper bound of LRRR efficiency as a function of time.

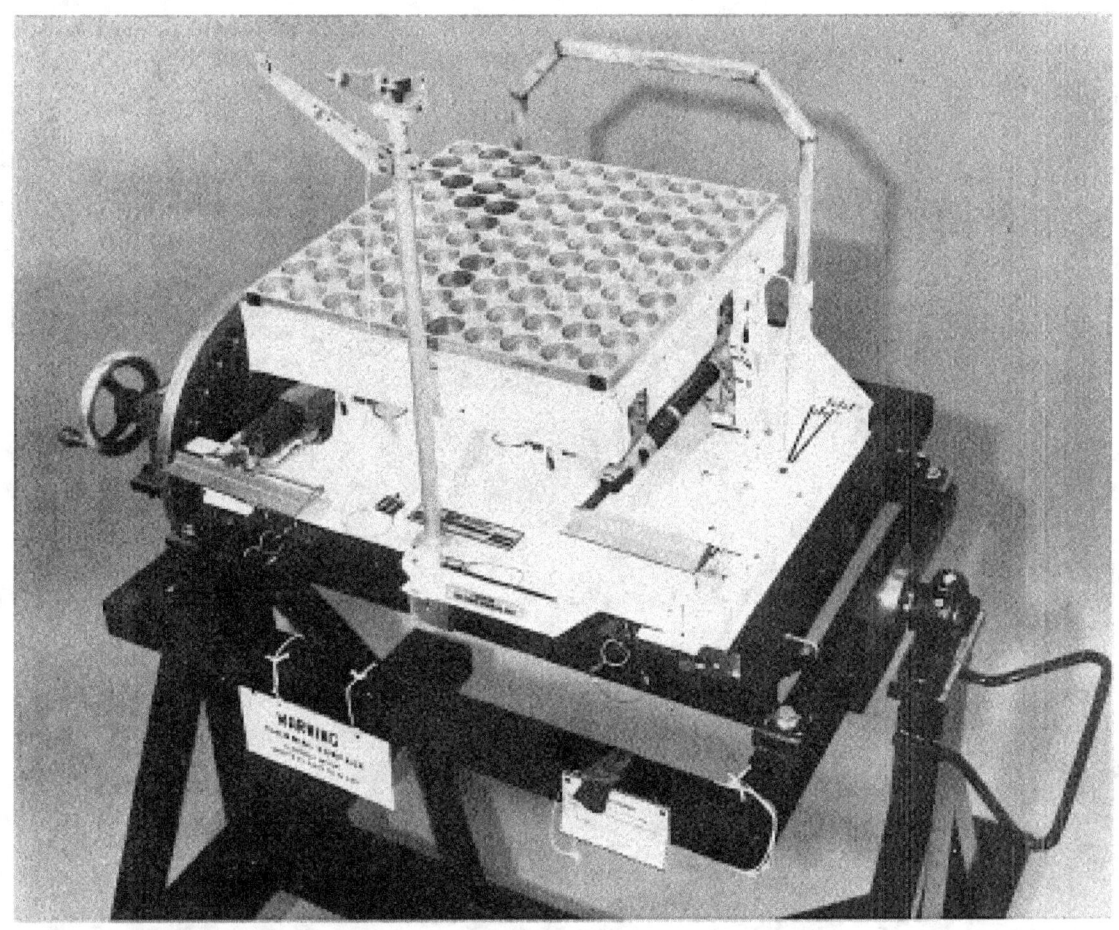

FIGURE 7-6.—The LRRR before stowage in the scientific equipment bay.

August 20, 1969 (immediately after lunar sunrise), show that the mounting (designed to minimize temperature gradients at these large Sun angles) has been successful. Observations of returns on September 3 and 4, 1969 (during lunar night), confirm the expectation that the reflector would perform at very low temperatures and that the differential thermal contractions of the mount and corners do not produce large strains. The survival of the reflectors throughout a lunar night has been demonstrated. Performance degradation, if any, caused by the presence of debris during lunar module (LM) ascent does not appear to be severe.

Ground Station Design

A single telescope can be used both as a transmitter and a receiver because the large diffraction pattern resulting from the 3.8-cm-diameter corner reflectors (the central spot has a diameter on the Earth of approximately 10 miles) allows for a velocity aberration displacement of approximately 1 mile without significant loss of signal. This was one of the major considerations in the design of the LRRR array discussed previously. The 2.5-sec light travel time between transmitting and receiving readily allows the mechanical insertion of a mirror that directs the returning photons collected by the telescope into a photomultiplier detector.

The present beam divergence of short-pulse, high-energy ruby lasers requires the use of a large aperture to recollimate the beam so that the beam is narrowed to match the divergence allowed by the atmospheric fluctuations of the

FIGURE 7-7. — The LRRR deployed on the Moon.

index of refraction — typically several seconds of arc. Astronomers refer to this effect as "seeing." The degree of atmospheric turbulence at an observatory at a given time thus determines the size of the laser beam on the lunar surface, producing a spread of approximately 2 km per sec of arc divergence.

Techniques for pointing such narrow beams to a specific location on the Moon were developed during the successful Surveyor 7 laser-beam-pointing tests (ref. 7-7). An argon-ion laser beam was brought to a focus in the telescope focal plane at the Moon-image spot that was chosen for illumination. When the laser beam filled the exit pupil of the telescope and matched the f-number, the collimated beam was projected to the selected location on the Moon and detected by the television camera on the Surveyor

7. For the Apollo laser ranging experiment, the beam is matched into the telescope by using a diverging lens, because it is not possible to focus high-power ruby laser beams in air without causing electrical breakdown. The direction of the projected beam is monitored by intercepting a small portion of the beam with corner reflectors mounted on the secondary mirror-support structure. These reflectors return the intercepted light in such a manner that the light is brought to focus, superimposed on the image of the Moon, at the spot to which the beam is being sent. A beam splitter coated with a highly damage-resistant, multilayer dielectric coating reflects the laser beam into the telescope and transmits the image of the Moon and the laser light intercepted by the telescope corner reflectors into the guiding system (ref. 7–8).

A view of the region of the Moon around Tranquility Base is shown in figure 7–8, which was taken through the guiding eyepiece of the McDonald Observatory 107-in. telescope. The reticle marks used to guide from craters are clearly shown, along with the intercepted laser light. In final alinement, the small circle is made to coincide with both Tranquility Base and the

FIGURE 7-9. — Lick Observatory, University of California, at Mount Hamilton, California.

FIGURE 7-10. — Lick Observatory 120-in. telescope dome.

FIGURE 7-8. — View through guiding eyepiece of the McDonald Observatory 107-in. telescope during process of alinement.

center of the laser spot. The large size of the laser spot is caused by imperfections in the telescope corner reflectors, not by beam divergence. The diameter of the reticle circle is approximately 3.6 seconds of arc. Many craters are not resolved because of poor seeing conditions at the time the photograph was made.

Figures 7–9 to 7–14 are views of the Lick Observatory, University of California, at Mount Hamilton, California; the Lick Observatory 120-in. telescope, the second largest in the world; the McDonald Observatory, University of Texas, at Mount Locke, Tex.; and the McDonald Observatory 107-in. telescope, the third largest in the world. Lasers were mounted in a stationary position at a coudé focus of each instrument.

The Lick Observatory participated in the acquisition phase of the experiment to increase

FIGURE 7-11. — Lick Observatory 120-in. telescope.

FIGURE 7-13. — McDonald Observatory 107-in. telescope.

FIGURE 7-12. — McDonald Observatory, University of Texas, at Mount Locke, Tex.

the probability of getting early returns because the weather and seeing there are generally excellent in the summer. Returns at the Lick Observatory, which is no longer in operation, were observed on August 1 and 3, 1969. The McDonald Observatory is equipped to make daily range measurements, weather permitting, for years. Returns were observed on August 20, September 3, September 4, and September 22, 1969, at the McDonald Observatory.

Observations at the Lick Observatory

System A

The ranging system at the Lick Observatory consisted of a giant-pulse, high-powered ruby laser (operated at the coudé focus) which was optically coupled through the 120-in. telescope and could be fired at 30-sec intervals. The angular diameter of the outgoing beam was approximately 2 seconds of arc and made a spot of light on the Moon approximately 2 miles in diameter. The return signal was detected by a photomultiplier that was mounted at the coudé focus behind a 10-second-of-arc field stop and a narrow (0.7 Å) filter, which were used to reduce the background illumination from the sunlit Moon. A time-delay generator (TDG), initiated by the firing of the laser, was used to activate the acquiring electronics approximately 2.5 sec (the Earth-Moon round-trip time for light) later. The

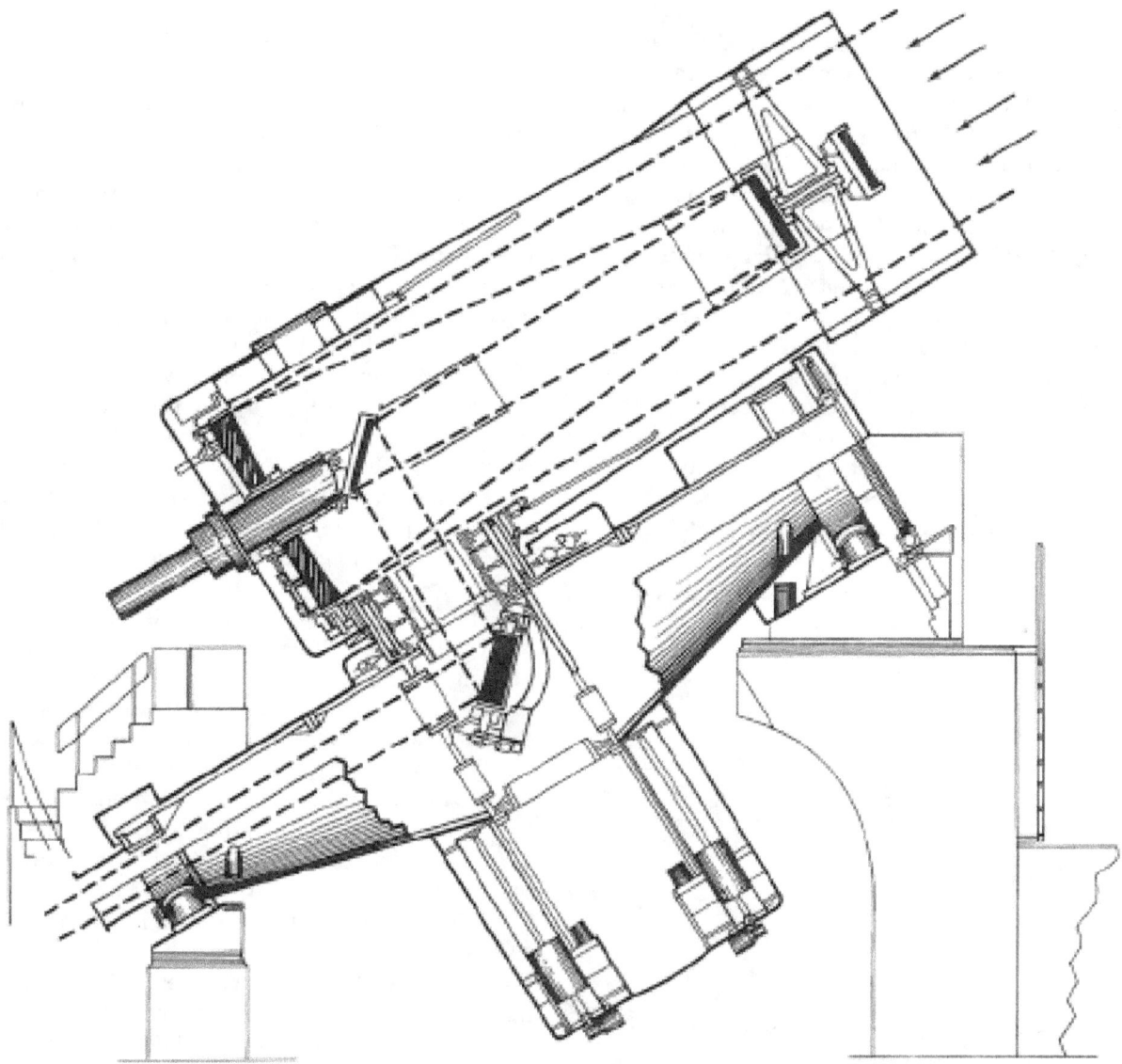

FIGURE 7-14.—Optical path in the McDonald Observatory 107-in. telescope.

delay generator was set for each shot by using the LE 16 ephemeris. (See the section of this report entitled "The Lunar Ephemeris: Predictions and Preliminary Results.")

Following the pulse produced by the TDG, the output pulses from the photomultiplier were channeled sequentially into 12 binary scalers. Each scaler channel had a dwell time that was adjustable from 0.25 to 4 μsec (ref. 7-8). The routing of the pulses to the scalers was such that a pulse arriving within 0.1 μsec of the end of a channel would also add a count to the following channel. The scalers then contained a quantized summary of the detector output for a short time interval centered on the expected arrival time of the reflected signal. After each scaler cycling following a laser firing, a small online computer read the contents and reset the scalers. The computer stored the accumulated count for each of the scalers and provided a printed output and a cathode-ray-tube display of the data.

Scattered sunlight from the lunar surface produced a random background that slowly filled the 12 time channels. Because the return from the retroreflector occurred with a predetermined delay, the channel that corresponded in time to the arrival of the signal accumulated data at a faster rate than the other 12 channels. Figure 7-15 illustrates this point by showing the way in which the data actually accumulated during run 18. Following acquisition on the night of August 1, 1969, 169 shots were fired. Range gate errors occurred on 27 shots, and 22 shots were fired with the telescope pointed away from the reflector. For the remaining 120 shots, approximately 100 above-background counts were received. These results represent a return expectation in excess of 80 percent and show that all parts of the experiment operated satisfactorily. Assuming a Poisson distribution of the recorded photoelectrons, the returns correspond to an average of 1.6 detectable photoelectrons per shot. This number is a lower limit to the true average because interference effects and guiding errors probably reduced the number of returns that were recorded. The strength of this signal and the lack of "spill" into adjacent channels clearly show that the signal did not come from the "natural" lunar surface, from which the return would be distributed over approximately 8 μsec. The timing of the trigger from the TDG relative to Mulholland's ephemeris was changed three times, and the channel widths were decreased from 2 to 1 μsec and then to 0.5 μsec. After each change, the signal appeared in the appropriate channel. The data from runs 10 to 21, the interval from the first acquisition to the close of operation, are shown in table 7-IV. Nine runs containing 162 shots were made before acquisition.

Figure 7-16 shows a plot of the data taken from table 7-IV. Runs 12 and 14 have not been plotted in figure 7-16 because errors in setting the TDG invalidated the timing. During run 19, the telescope was not pointed at the reflector, and no returns were seen. Because of a fortuitous splitting of the return between two channels on two runs, an effective timing precision of 0.1 μsec was achieved. This precision is equivalent to a range error of approximately 15 m. In figure 7-16 an apparent drift in the time the returns were detected, relative to Mulholland's predictions, is shown. The drift was caused by the 120-in. telescope being located approximately 524 m east of the locations given for the Lick Observatory in the American Ephemeris and Nautical Almanac. A curve showing this correction to the original ephemeris is given in figure 7-16. Translation of this curve along the ordinate is allowable, and the amount of time gives the difference between the observed range and the predicted range. When the correct coordinates are used, the observations agree with the predicted curve.

System B

The primary function of acquisition system B at the Lick Observatory was to locate the deployed LRRR in range and position. System B was not intended to satisfy the long-term objectives of the lunar ranging experiment; however,

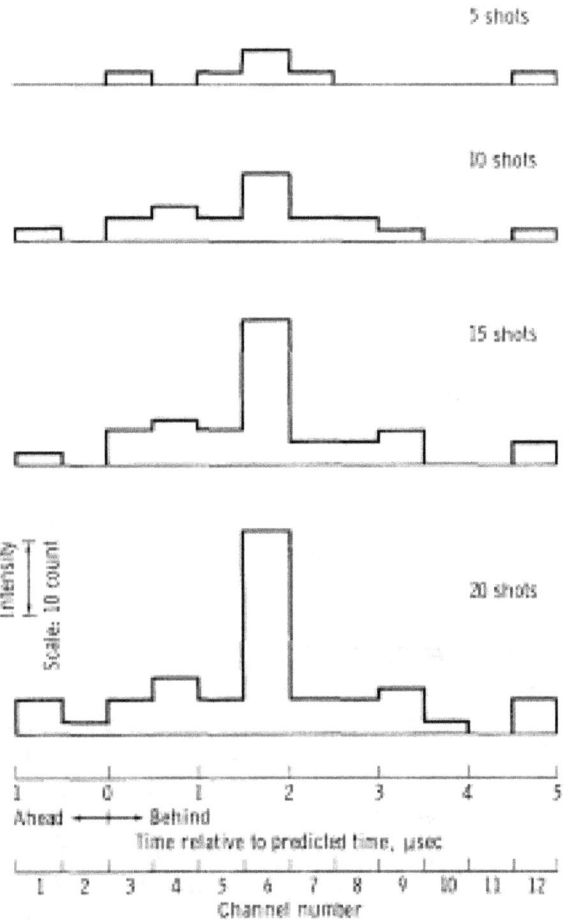

FIGURE 7-15.— Histogram showing the growth of the retroreflected signal in channel 6.

TABLE 7–IV. *Log of observation data showing acquisition*

Run	Total counts Channel												No. of shots	u.t. for middle of run, hr:min	Channel width, μsec	Time of first channel[a] μsec
	1	2	3	4	5	6	7	8	9	10	11	12				
10	12	8	16	18	12	14	10	17	13	[b]27	12	12	20	10:21	2.0	−20
11	12	12	12	11	11	6	13	11	14	[b]28	10	14	14	10:32	2.0	−20
12	(c)	(c)	(c)	(c)	(c)	(c)	(c)	(c)	(c)	(c)	(c)	(c)	16	—	2.0	−10
13	13	8	8	12	7	[b]18	11	5	6	7	8	12	13	11:04	2.0	−10
14	(c)	(c)	(c)	(c)	(c)	(c)	(c)	(c)	(c)	(c)	(c)	(c)	6	—	1.0	−10
[d]15	4	3	3	5	4	[b]17	6	8	10	5	6	8	18	11:23	1.0	−5
16	1	1	2	2	[b]6	3	3	1	2	1	3	2	10	11:36	.5	−1
17	6	3	4	2	[b]11	[b]9	2	7	2	4	2	5	16	11:45	.5	−1
18	3	1	3	5	3	[b]19	3	3	4	1	0	4	22	12:03	.5	−1
[e]19	3	3	3	10	4	3	5	2	5	5	8	5	22	12:19	.5	−1
[f]20	2	1	1	0	3	4	[b]6	2	4	2	2	4	10	12:23	.5	−1
21	5	2	2	3	2	1	[b]12	[b]11	3	4	5	2	22	12:45	.5	−1

[a] With respect to ephemeris predictions.
[b] Channel in which return was expected.
[c] Range-gate errors invalidated data.
[d] Data from three shots with erroneous range gates deleted from tabulation.
[e] Telescope pointed 16 km south of reflector.
[f] Thin clouds noted near Moon.

System B was designed to be a sensitive, moderately precise, semiautomatic, high-repetition rate system using existing off-the-shelf equipment wherever possible.

System B was composed of several essentially independent subsystems. These subsystems were the laser-transmitter/power-supply assembly, the receiver-detector ranging system, the range gate generator and data control and recorder system, and the time standard system.

The laser transmitter uses oscillator and amplifier heads employing a conventional rotating-prism, bleachable-absorber, "Q-switch" mechanism. The oscillator and amplifier heads are identical and use 110-mm-long, 15-mm-diameter Brewster-Brewster rubies. An existing laser system that was modified for this program was capable of operation at 10 J at 3-sec intervals with a pulse width of 60 to 80 nsec.

For operation at the Lick Observatory, modification of the laser system required the incorporation of all normal transmit-receive ranging functions and of a boresight capability onto the laser case. An optical-mechanical assembly (fig. 7–17) was designed to be attached to the laser case. The exiting laser beam passed through a beam splitter, which was oriented at Brewster's angle for minimum reflection of horizontally polarized light. The small fraction of light scattered or reflected from the beam splitter was detected by an FW 114A biplanar photodiode. The output of the photodiode was used for the range measurement initiation pulse and for monitoring the operation of the laser. The beam splitter was also used to couple a vertically polarized helium-neon laser beam along the same axis as the ruby beam. A helium-neon laser was mounted parallel to the ruby laser, and the resultant beam was expanded by a small autocollimator and then reflected by a mirror mounted parallel to the beam splitter. The reflected beam was then rereflected by the beam splitter along the laser axis. The laser produced an elliptical cross-section beam that was corrected and expanded to 50 mm in diameter by the Brewster entrance prism telescope. The 50-mm exit beam was reflected at right angles onto a 50-mm-aperture F39 lens that was positioned at the appropriate place with respect to the coudé focus of the 120-in. telescope optical system.

The returning energy follows essentially the same path as the transmitted energy. A flip mirror, actuated by a small solenoid, reflects the returning signal after the signal is passed through the correcting telescope to a field-limiting lens and aperture. The aperture in the field-limiting

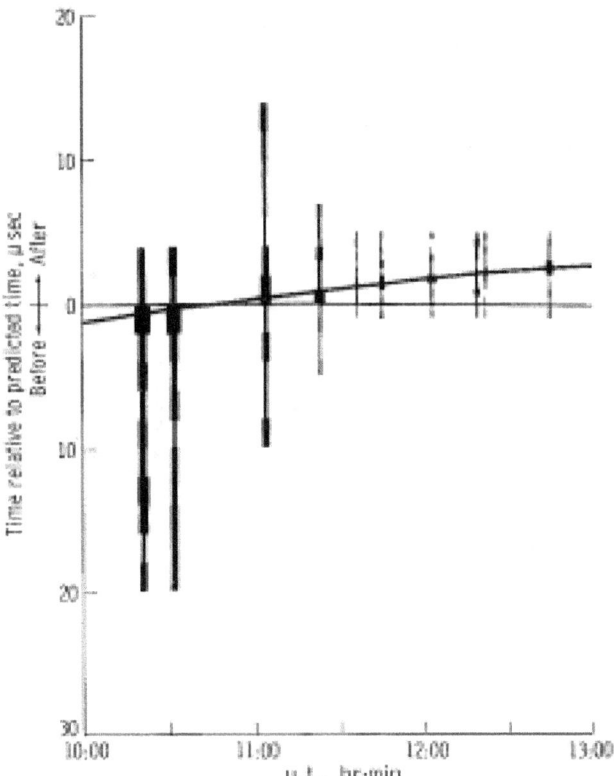

FIGURE 7-16. — Three-dimensional figure showing the time of the run (abscissa), the range window during which the equipment was open for receiving data (ordinate), and the number of counts in each channel (width of the bar). It is clear that in each run in which returns were expected they were seen in the correct channel. During run 19 (third from right-hand end), the telescope was pointed away from the reflector, and no returns were seen.

system is sized for a 10-second-of-arc field of view in space. Light passing through the aperture is collimated and passed through a 2.7-Å bandpass filter. The light is divided into two equal beams by a dividing prism; then, the beams of light enter the enhancement prism on each of the two photomultipliers.

Two detectors (operating in coincidence) and a range gate were used to eliminate as much extraneous background noise as possible. Type 56 TVP photomultipliers used enhancement prisms to allow multiple reflection of the entrance light from the photocathode. The output of each photomultiplier tube was discriminated and shaped with an EGG T105/N dual discriminator. Because theoretical considerations indicate that only a few photoelectrons occur for each transmitted pulse, the discriminator was used to stretch each detected pulse by the length of the transmitted pulse. A coincidence overlap equal to the transmitted pulse duration was necessary because, lacking discrimination, a photoelectron from each photomultiplier tube could be related to any time within the transmitted pulse duration. The outputs of the discriminators were AND'ed in the EGG C102B/N coincidence module.

The actual range measurement (fig. 7–18) was made with a 1-nsec time-interval counter. The time-interval counter was started by the photodiode output each time the laser transmitter operated. The discriminators and the stop channel of the time-interval unit were disabled by the range gate until the expected time of arrival of the reflected laser pulse. If both photomultiplier tubes detected a photoelectron within the coincidence resolving time of the EGG C102B/N, the output would stop the time-interval unit.

The photodiode signal was used in two other measurements. First, the signal was used to stop a 100-μsec-resolution time-interval unit that was started by the "ontime" timing signal from the real-time clock. This measurement determined the time, to the nearest 100 μsec, of the range measurement. The pulse was also converted, monitoring the overall performance of the laser during the operation.

System B was controlled and operated by a special digital logic assembly. This device (fig. 7–19) generated laser fire signals and the range gate window, sampled all measurement devices for information, and recorded the information. One of the most important aspects of the operation is the generation of the range gate window, based on knowledge of where the retroreflectors should be. Furthermore, because this information may not be as accurate as required, it is necessary to change the information essentially in real time. The range gate window was generated with an externally, as well as manually, programed delay pulse generator. The programed input for the delay pulse generator was obtained from a lunar prediction drive tape containing the expected round-trip time interval to the lunar surface for every 3 sec of time. A digital comparison between the command time on the tape and real time was made to maintain the tape in

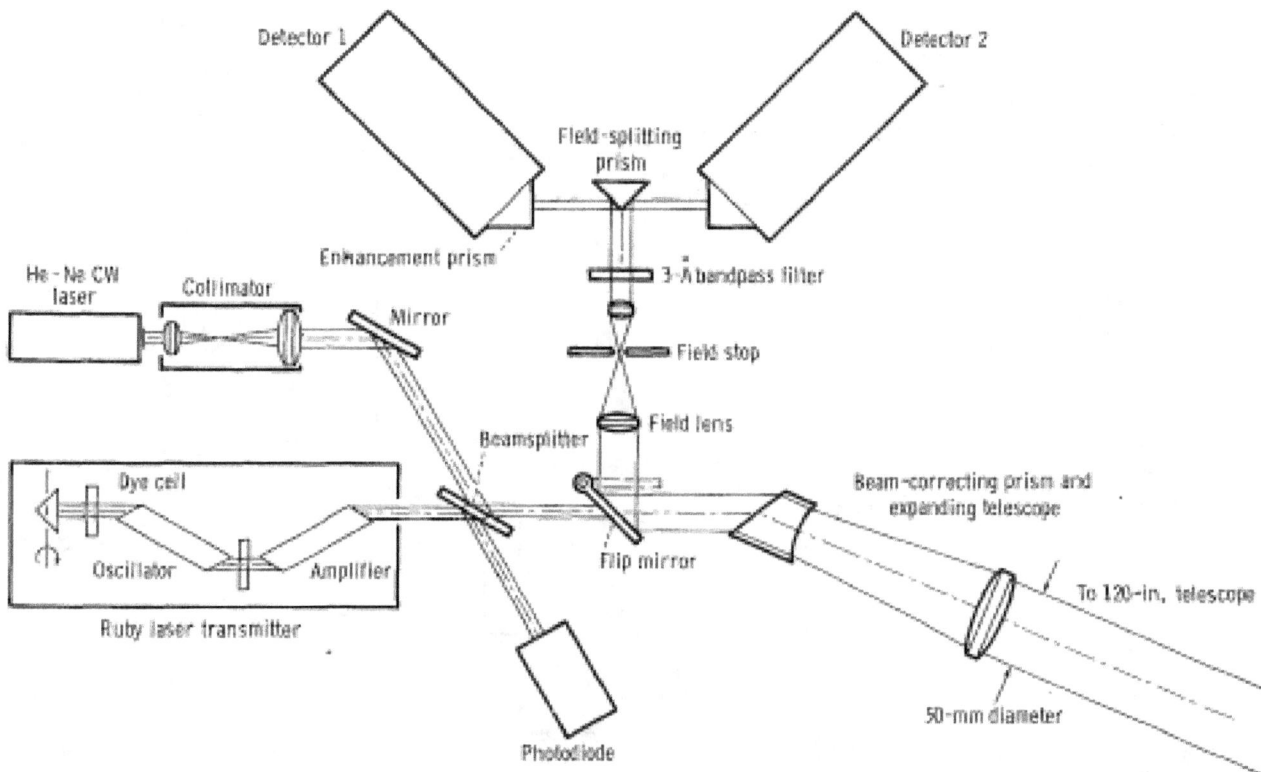

FIGURE 7-17. — Optical schematic of system B at Lick Observatory.

synchronization with real time. The range information on the tape was then stored and transferred to the delay pulse generator prior to each laser firing. The laser firing started the delay pulse generator; and after the predicted delay, a pulse initiated the range-gate-window pulse generator. The range-gate-window length was variable and was used to enable the detector discriminators and the range time-interval unit at the time a return signal was expected.

The measured data were recorded with a multiplexing data gate and standard line printer. The control section generated print commands for each series of measurements at the correct time during the 3-sec cycle period. The control section obtained timing signals from the time standard rack (fig. 7–20). Real time was maintained to an accuracy of ± 10 μsec during the operation, by comparing signals from the Loran C chain with the real-time clock using the Loran C synchronization generator.

The system was operated for a period of 1 hr and 45 min on the morning of August 3, 1969, during which the laser was fired 1230 times.

The total number of apparent measured ranges during this period was 98. The low number of returns relative to the number of transmitted pulses is related to a slight (2 sec of arc) misalinement in the detector optical system and to the failure of the delay pulse generator to operate in the externally programed mode. Furthermore, because returns were not immediately recognized, an angle search was made over a longer time period.

The measured range time could have three possible sources: a return from the retroreflector, a return from the lunar surface, or random noise coincidence from reflected sunlight and background. Statistically, 35 to 50 noise coincidences and approximately 30 lunar surface ranges would be expected in 1200 firings. Because the numbers agree, within acceptable limits, with the total number measured, it is not obvious that returns from the retroreflectors were measured. However, any returns from the retroreflectors should fall within the precision of the system or within 100 nsec with respect to the true range to the package.

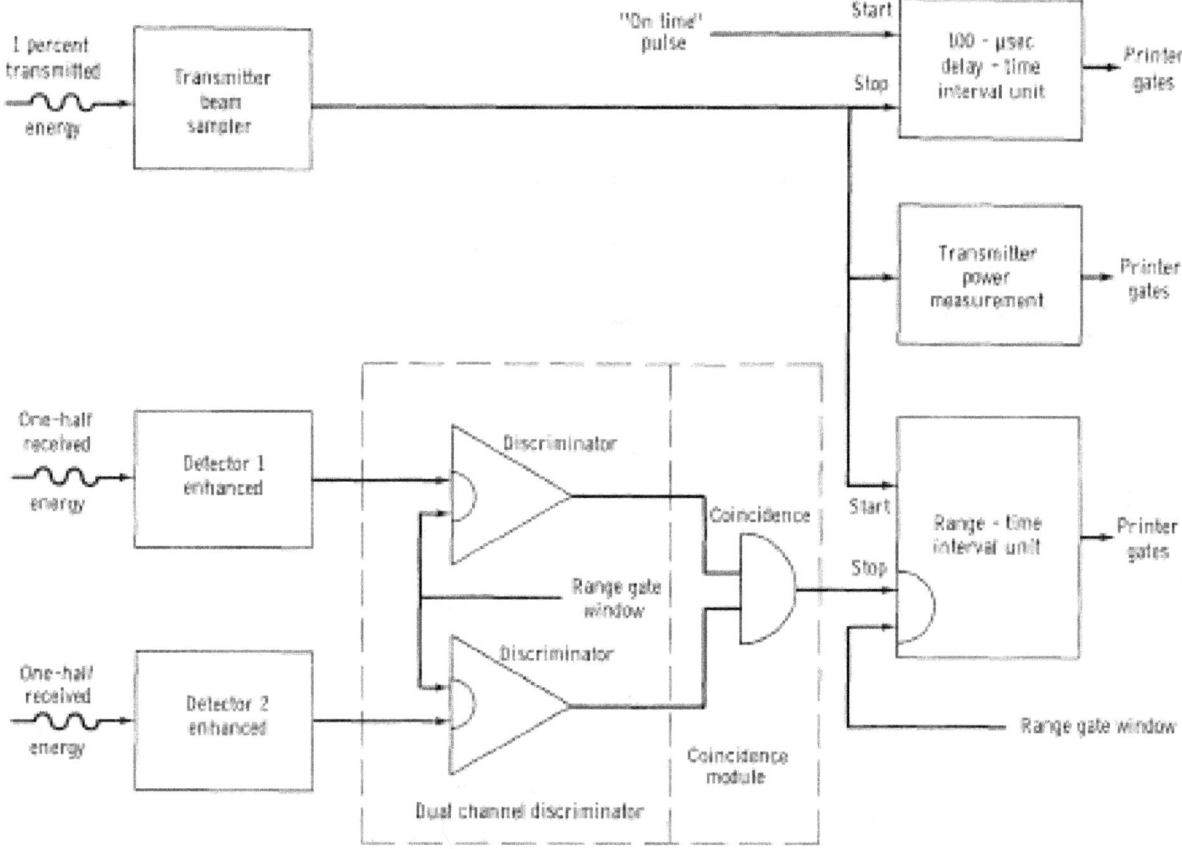

FIGURE 7-18. — Receiver and ranging section.

The observed range time measurements were compared arithmetically with the predicted range time and then linearly corrected because of a known parallax error. The parallax error occurred because a different site was used for the range predictions. The linear correction was performed by taking the initial range residuals and fitting (in a least squares sense) a first-order polynomial to the residuals. Then, only those residuals occurring within a certain range of the polynomial were used to redefine the polynomial. This technique was used successively three times to generate a polynomial about which 11 of the measured ranges had a root mean square of 45 nsec. Furthermore, because of the parallax error, the coefficients of the polynomial agree with the expected deviations from the predicted range.

The range residuals corrected to this linear equation were used to construct a histogram (fig. 7–21) with 100-nsec intervals from −5 to +5 μsec. The data lying outside of the ±5 μsec were not displayed because no interval contained more than one point and because no significant bunching was observed. The histogram clearly shows a central peak that cannot be supported by any statistical interpretation other than one assuming returns from the retroreflector.

Observations at the McDonald Observatory

The laser in use presently is a custom-built two-stage pockels-cell switched ruby system. Typical operating conditions were as follows:

(1) Energy: 7 J
(2) Pulse width: 20 nsec
(3) Beam divergence: 2.4 mrad (measured at full energy points)
(4) Repetition rate: once every 6 sec
(5) Wavelength: 6943.0 ± 0.2 Å at 70° F
(6) Amplifier rod diameter: 0.75 in.

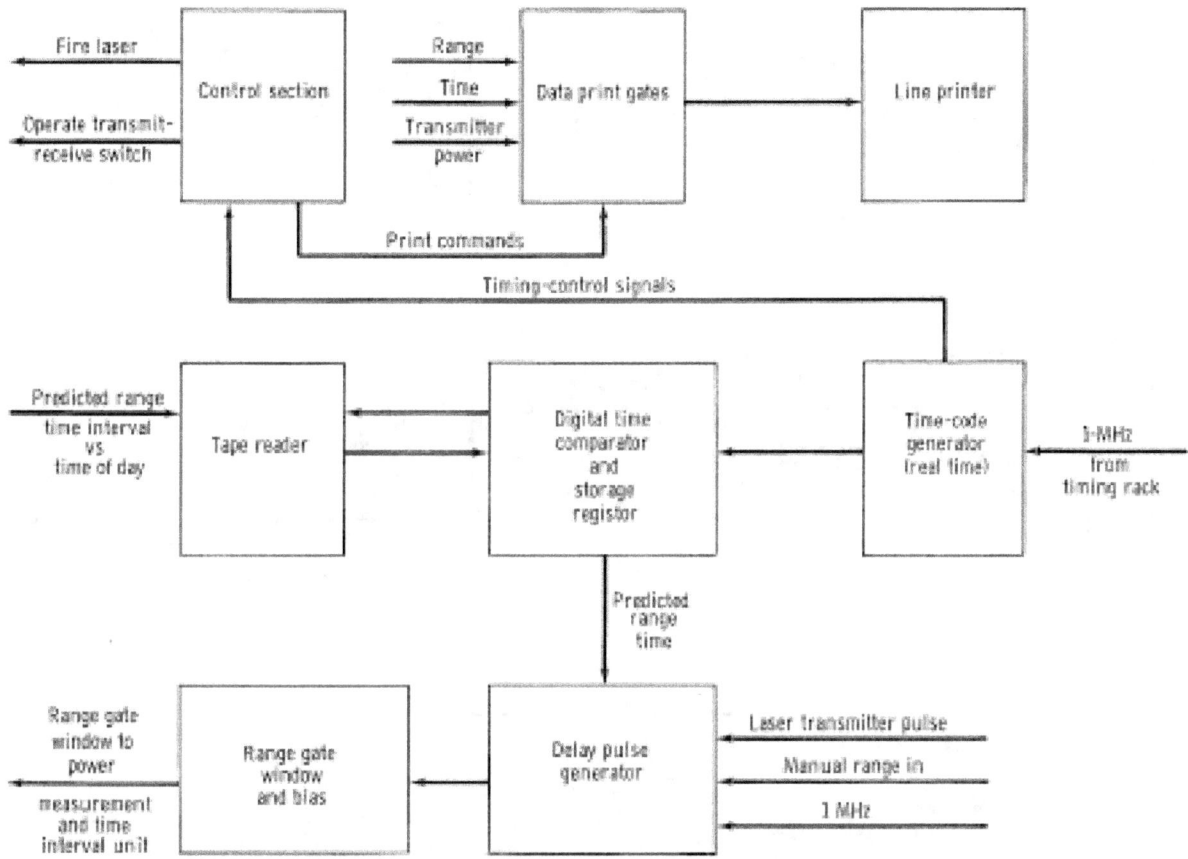

FIGURE 7-19. — Range gate generator and data control.

The detector package contained a photomultiplier that had a measured quantum efficiency of 5 percent at 6943 angstroms and a dark current of 80 000 counts per second. When cooled by dry ice (as during ranging operations), the dark current was 10 000 counts per second. Spectral filters with widths of 3 and 0.7 Å were available. Both filters were temperature controlled. Pinholes restricting the field of view of the telescope to 6" or 9" were commonly used. An air-driven protective shutter was closed during the time of laser firing, opening for approximately 1 sec around the time for receiving returns. The net efficiency of the whole receiver, ratio of photoelectrons produced to photons entering the telescope aperture (with a 3-Å filter), including telescope optics, was measured to be 0.5 percent, using starlight from Vega.

The block diagram of the timing electronics used during the acquisition period is shown in figure 7-22. The electronics consist of a multistop time-to-pulse-height converter (MSTPHC) for coarse range search covering an interval of 30 μsec with 0.5-μsec bins (ref. 7-8) in addition to the core circuits forming part of the intended subnanosecond timing system. The initial and final vernier circuits of this system were not in use. The range prediction provided by J. D. Mulholland was recorded on magnetic tape at 6-sec intervals. The online computer read the range prediction, set the range gate TDG, and fired the laser within 1 μsec of the integral 6-sec epoch. The TDG activated the MSTPHC, triggered a slow-sweep oscilloscope (the display being recorded on photographic film) and a fast-sweep oscilloscope (recorded on Polaroid), and activated a 10-μsec gate into the time-interval meter (TIM). The computer read the number of counts in the TIM and calculated the difference between this reading and the range prediction, printing out

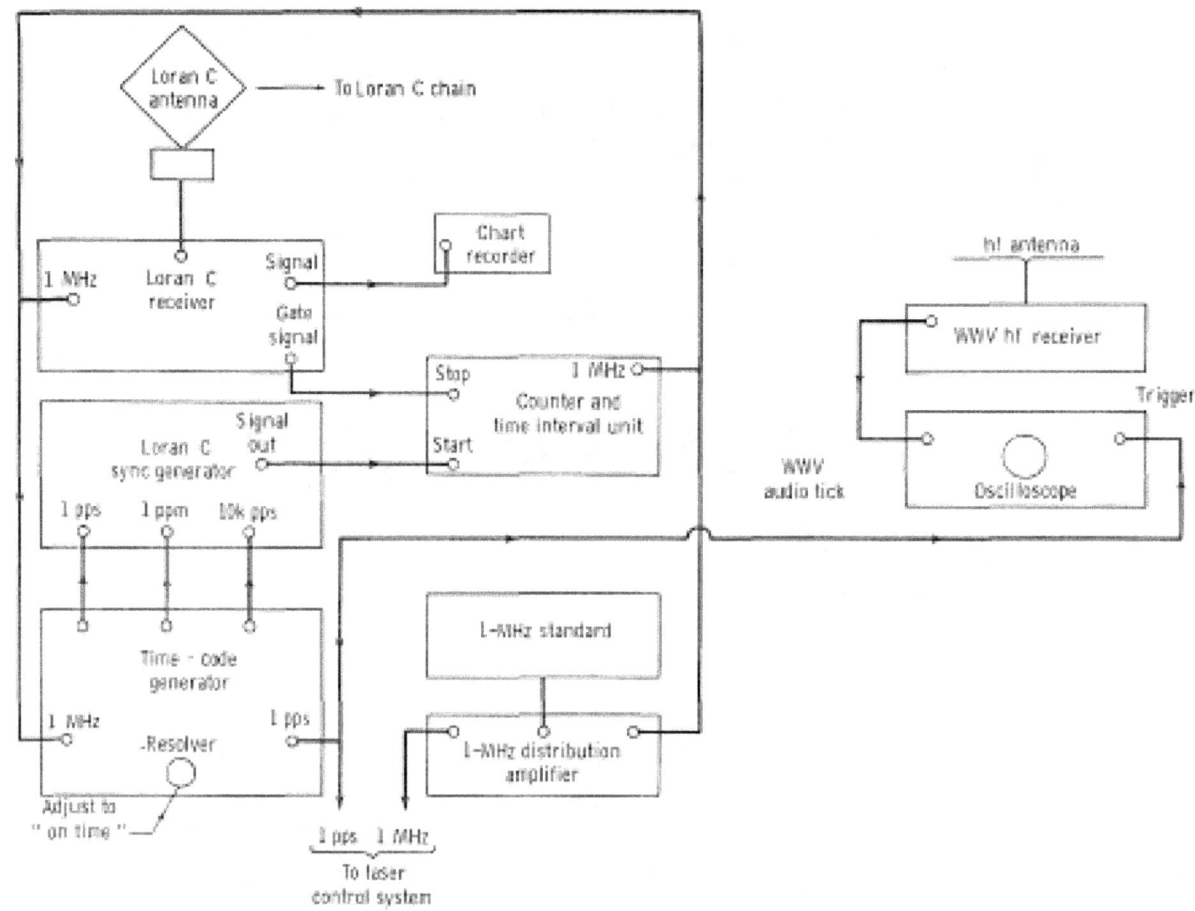

FIGURE 7-20. — Block diagram, time standard system for system B at Lick Observatory.

this difference on the teletypewriter to the nearest nanosecond. The MSTPHC range accuracy depended on the TDG, whereas the TIM range accuracy is entirely independent of the TDG.

The first high-confidence-level return was recorded by the MSTPHC for a 50-shot run at approximately 2:50 Greenwich mean time (G.m.t.) on August 20, 1969 (fig. 7-23). A part of the corresponding TIM printouts is displayed as a histogram in figure 7-24. Here, the origin of the time axis is at the predicted range. The lower histogram shows a portion of the printouts for a 50-shot run taken a few minutes later in which a 5-μsec internal delay was introduced. (This delay has been subtracted in the drawing.) Noise scans in which the laser was fired into a calorimeter displayed no buildup. Four other scans recording signals were made in the 50 min before the Moon sank too low in the sky. Operation earlier in the night had been prevented by cloud cover.

The randomness of the difference between the TIM printout and the ephemeris prediction enabled a statistical reduction of the data even without the vernier circuits designed to interpolate between the 50-nsec digital intervals. The result is a measured round-trip travel time in excess of the Mulholland prediction by 127±15 nsec of time at 3:00 G.m.t., August 20, 1969, from the intersection of the declination and polar axes of the 107-in. telescope. The uncertainty corresponds to ±2.5 m in one-way distance.

Return signals were again recorded on September 3 and 4, 1969, with equivalent uncertainty. Round-trip travel times were also shown in excess of the prediction by 497±15 nsec on September 3, 1969, at 11:10 G.m.t. and by

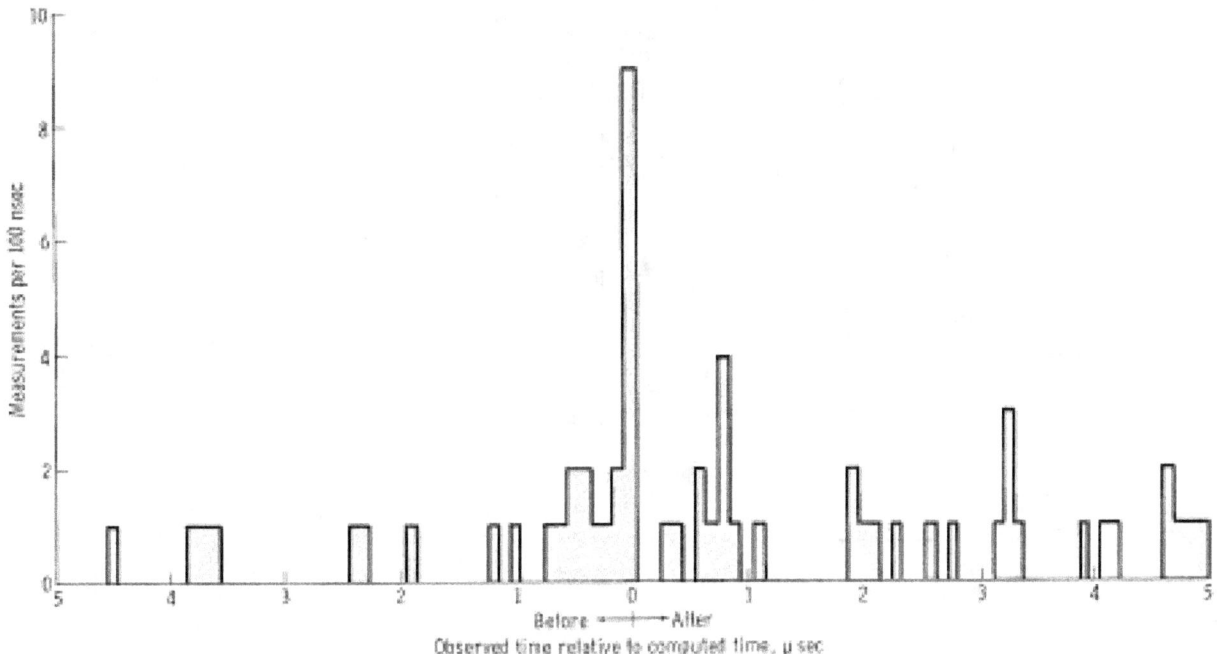

FIGURE 7-21.— Range residuals from system B at Lick Observatory.

797±24 nsec on September 4, 1969, at 10:10 G.m.t. During these observations, Tranquility Base was in darkness, and the computer-controlled drive of the telescope was used successfully to offset from visible craters and track the reflector.

The Lunar Ephemeris: Predictions and Preliminary Results

The fundamental input to the calculation of predictions for the LRRR is the lunar ephemeris, which gives the geocentric position and velocity of the lunar center of mass. An ephemeris is being used that was developed at the Jet Propulsion Laboratory (JPL) and is designated LE 16. This ephemeris is believed to be far superior in the range coordinate to any other extant ephemeris. The available observational evidence is meager but supports this belief. The modeling of the topocentric effects relating to the motions of the observatory and the reflector about the centers of mass of the respective bodies is complete and is similar to that used in JPL spacecraft tracking programs. Universal time (u.t.) 1 is modeled and extrapolated by polynomials fit to the instantaneous determinations by the U.S. Naval Observatory Time Service. The Koziel-Mitielski model is currently being used for the lunar librations.

Coordinates of the telescope and the reflector package are input variables to the prediction program. Reflector coordinates presently used are those derived at the NASA Manned Spacecraft Center from spacecraft tracking of the lunar module, since this dynamic determination is essentially the inverse of the predictive problem and is, thus, more compatible than the selenographic determinations.

It was anticipated that the topographic modeling would be the primary error source in the earliest phase of ranging operations. This belief was based upon the indications from command and service module (CSM) tracking of Apollo 8 and Apollo 10, both of which indicated LE 16 ephemeris errors of 40 to 50 m (0.3 μsec), and is based on the knowledge that various estimates of the selenocentric distances of surface locations disagreed by perhaps 2 km for a specific region. The decision to use the dynamic determinations of the location simplified the real-time processes but did not relieve the radial distance uncertainty until laser acquisition was an accom-

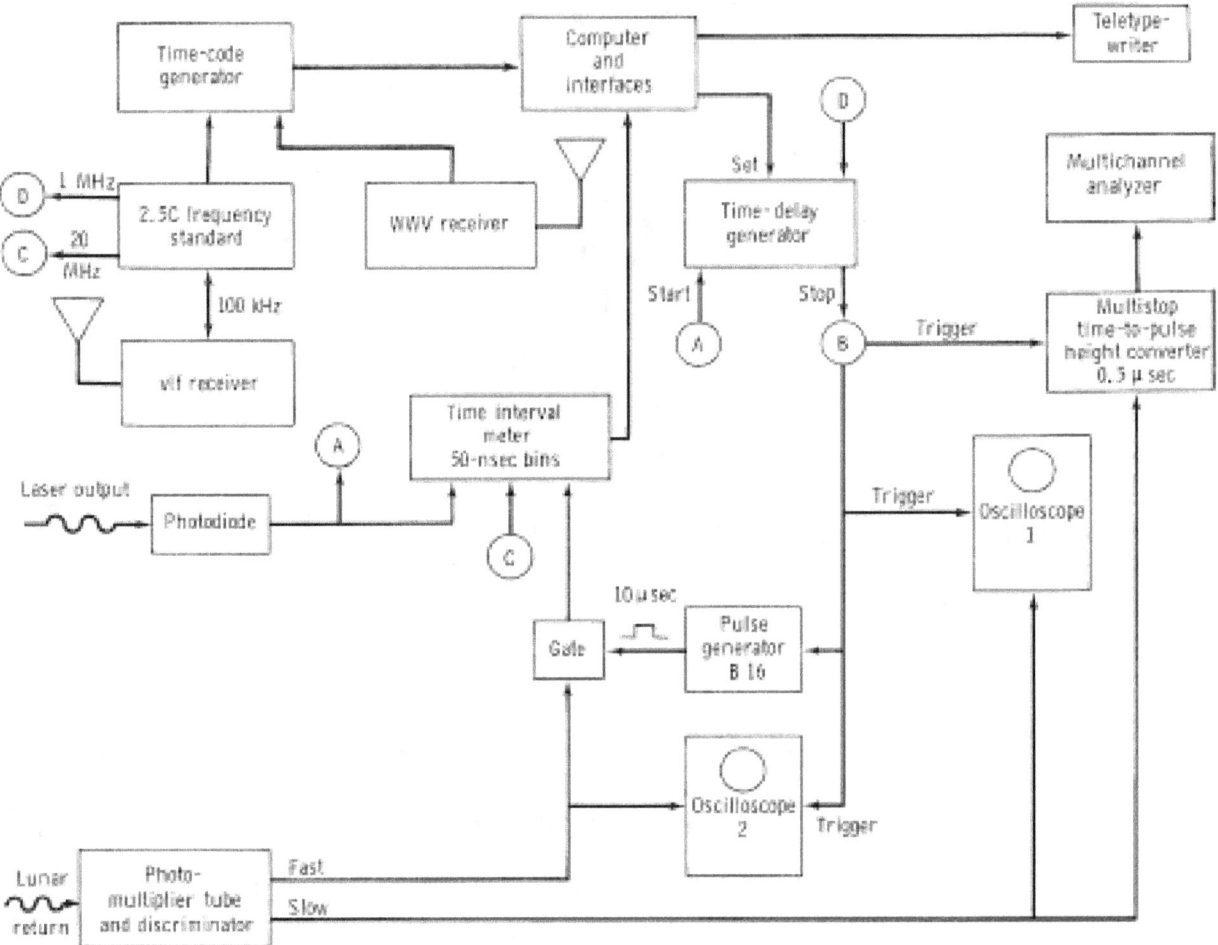

FIGURE 7-22.—Block diagram of McDonald Observatory acquisition and early measurement phase electronics.

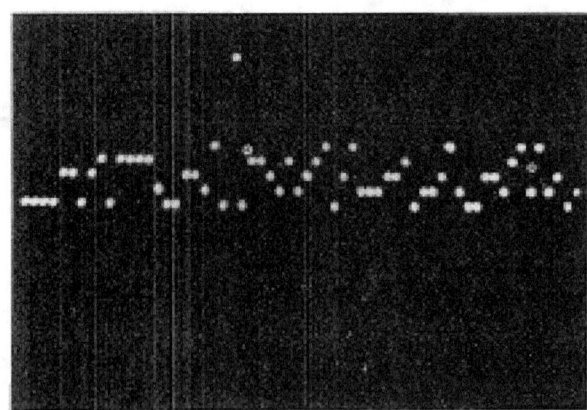

FIGURE 7-23.—MSTPHC display of McDonald Observatory acquisition.

plished fact. As the current "best estimate" of the landing site shifted, predicted ranges were affected by approximately 4 km (28 μsec). Site coordinates for Tranquility Base provided by the NASA Manned Spacecraft Center on July 22, 1969, are currently being used.

Tracking information is available on both the CSM and the lunar module during lunar surface operations. The CSM data indicate an ephemeris error not greater than 50 m. Computations of predictions for comparison with the LM ranging data, using the coordinates mentioned previously, show residuals of approximately 2 km, with a drift of 0.3 km in 15 min. The explanation of this anomaly is not yet known, but it seems to be associated with the tracking station location.

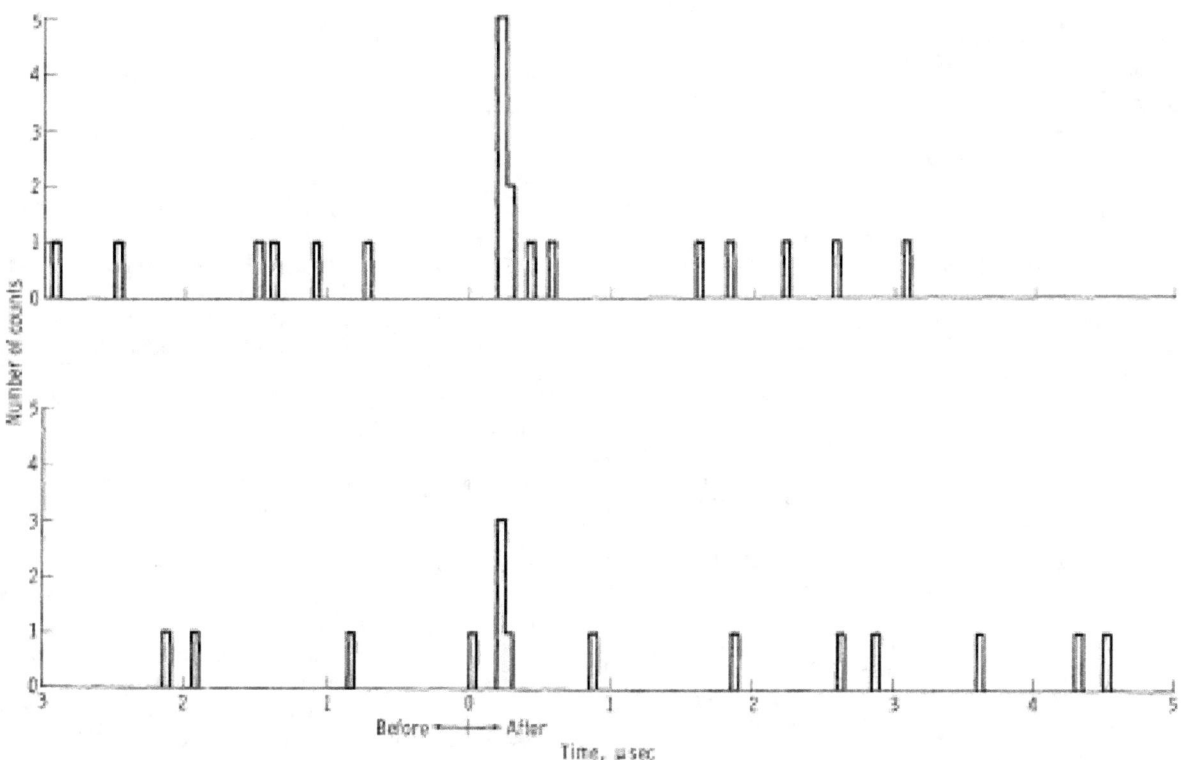

FIGURE 7-24. — Histogram of single-stop TIM readings on two successive 50-shot runs on August 20, 1969, at the McDonald Observatory.

Initial acquisition of the LRRR was accomplished with predictions based on the coordinates of the Lick Observatory, as published in the American Ephemeris and Nautical Almanac. The Lick Observatory 120-in. telescope used in the experiment is actually a distance of approximately 1800 ft from that location, causing drift in the observation residuals. The result of introducing the proper telescope coordinates into the computations is shown in figure 7–25. The residuals relative to the new telescope coordinates appear to be given within 1 μsec. Figure 7–25 is provisional and subject to later refinement. Within the limitations of figure 7–25, these data are consistent with the subsequent observations at McDonald Observatory.

Although it is premature to discuss the data from the standpoint of any meaningful application, one aspect invites speculation and preliminary inference. The three indirect pseudo-observations (range biases on CSM tracking data)

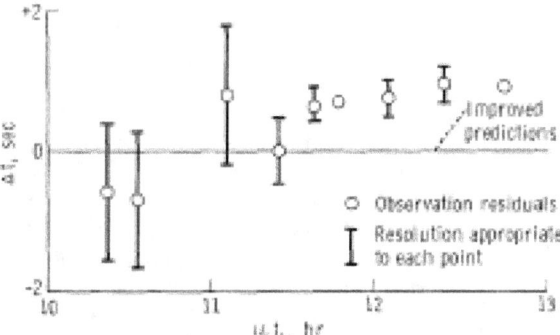

FIGURE 7-25. — Result of using adjusted telescope coordinates for the Lick Observatory. The data are shown for August 1, 1969.

and the first few laser acquisitions yield positive observation residuals. If no observation selection effect is involved, this may be an indication that the ephemeris scaling factor and hence the lunar mean distance requires an adjustment in the seventh place. Such a result would not be surprising because the uncertainty in the scaling factor is five parts in 10^7 (ref. 7–10), based on the results of Mariner spacecraft tracking.

References

7-1. ALLEY, C. O.; BENDER, P. L.; DICKE, R. H.; FALLER, J. E.; FRANKEN, P. H.; PLOTKIN, H. H.; and WILKINSON, D. T.: Optical Radar Using a Corner Reflector on the Moon. J. Geophys. Res., vol. 70, May 1965, p. 2267.

7-2. ALLEY, C. O.; and BENDER, P. L.: Information Obtainable from Laser Range Measurements to a Lunar Corner Reflector. Symposium No. 32 of the International Astronomical Union on Continental Drift, Secular Motion of the Pole, and Rotation of the Earth, W. Markowitz and B. Guinot, eds., D. Reidel Publishing Co. (Dordrecht, Holland), 1968.

7-3. ALLEY, C. O.; BENDER, P. L.; CURRIE, D. G.; DICKE, R. H.; and FALLER, J. E.: Some Implications for Physics and Geophysics of Laser Range Measurements from Earth to a Lunar Retro-Reflector. Proceedings of NATO Advanced Study Institute on the Application of Modern Physics to the Earth and Planetary Interiors, S. K. Runcorn, ed., John Wiley & Sons (London), 1969.

7-4. MACDONALD, G. J. F.: Implications for Geophysics of the Precise Measurement of the Earth's Rotation. Science, 1967, pp. 157 and 204-205.

7-5. ALLEY, C. O.; ET AL.: Laser Ranging to Optical Retro-Reflectors on the Moon. Univ. of Maryland proposal to NASA, Dec. 13, 1965 (Rev. Feb. 11, 1966); Appendix VII – Design and Testing of Lunar Retro-Reflecting Systems by J. E. Faller.

7-6. ALLEY, C. O.; CHANG, R. F.; CURRIE, D. G.; FALLER, J. E.; ET AL.: Confirmation of Predicted Performance of Solid Fused Silica Optical Corner Reflectors in Simulated Lunar Environment. Univ. of Maryland Progress Report to NASA, Oct. 10, 1966.

7-7. ALLEY, C. O.; and CURRIE, D. G.: Laser Beam Pointing Tests. Surveyor Project Final Report, Part 2, Science Results. Jet Propulsion Laboratory NASA Tech. Rept. 32-1265, June 15, 1968.

7-8. SILVERBERG, E. C.: An Inexpensive Multichannel Scaler With Channel Widths of Less than One Microsecond. Rev. of Sci. Inst., Oct. 1969.

7-9. CURRIE, D. G.: Some Comments on Laser Ranging Retro-Reflector Ground Stations. Univ. of Maryland Tech. Rept. 856, Sept. 4, 1968.

7-10. MELBOURNE, W. G.; MULHOLLAND, J. D.; ET AL.: Constants and Related Information for Astrodynamic Calculations, 1968. Jet Propulsion Laboratory Tech. Rept. 32-1306, July 15, 1968.

ACKNOWLEDGMENTS

Responsibility for the design and conduct of the experiment has rested with the following groups: Principal Investigator, C. O. Alley (University of Maryland); Co-Investigators, P. L. Bender (National Bureau of Standards), R. H. Dicke (Princeton University), J. E. Faller (Wesleyan University), W. M. Kaula (University of California at Los Angeles), G. J. F. MacDonald (University of California at Santa Barbara), J. D. Mulholland (JPL), H. H. Plotkin (NASA Goddard Space Flight Center), and D. T. Wilkinson (Princeton University); Participating Scientists, W. Carrion (NASA Goddard Space Flight Center), R. F. Chang (University of Maryland), D. G. Currie (University of Maryland), and S. K. Poultney (University of Maryland).

The following people are responsible for the section of this report entitled "Observations at the Lick Observatory": James Faller and Irwin Winer (Wesleyan University); Walter Carrion, Tom Johnson, and Paul Spadin (NASA Goddard Space Flight Center); and Lloyd Robinson, E. Joseph Wampler, and Donald Wieber (Lick Observatory, University of California). These authors of the "Observations at the Lick Observatory" section wish to acknowledge the efforts of Norman Anderson (Berkeley Space Science Laboratory); Harold Adams, Raymond Greeby, Neal Jern, Terrance Ricketts, and William Stine (Lick Observatory); Barry Turnrose, Steve Moody, Tom Giuffrida, Dick Plumb, and Tuck Stebbins (Wesleyan University); and James MacFarlane, Bill Schaefer, Richard Chabot, James Hitt, and Robert Anderson (NASA Goddard Space Flight Center). Support from funds to the Lick Observatory by NASA grant NAS5-10752 and National Science Foundation grant GP 6310, from funds to Wesleyan by NASA Headquarters grant NGR-07-006-005, and from in-house funds used by Goddard personnel is acknowledged.

The people who are responsible for the section of this report entitled "Observations at the McDonald Observatory" include D. G. Currie, S. K. Poultney, C. O. Alley, E. Silverberg, C. Steggerda, J. Mullendore, and J. Rayner (University of Maryland); H. H. Plotkin and W. Williams (NASA Goddard Space Flight Center); and Brian Warner, Harvey Richardson, and B. Bopp (McDonald Observatory). The authors of this section wish to acknowledge the efforts of Harlan Smith, Charles Jenkins, Johnny Floyd, Dave Dittmar, Mike McCants, and Don Wells (McDonald Observatory); Faust Meraldi, Norris Baldwin, Charles Whitted, and Harry Kriemelmeyer (University of Maryland); and Jim Poland, Peter Minott, Cal Rossey, Jim Fitzgerald, Walter Carrion, Mike Fitzmaurice, and Herb Richard (NASA Goddard Space Flight Center).

Support for the activities at the McDonald Observatory from NASA Grant NGR 21-002-109 to the University of Maryland and from in-house funds at the Goddard Space Flight Center is acknowledged.

J. D. Mulholland (JPL) is responsible for the section entitled "The Lunar Ephemeris."

8. The Solar-Wind Composition Experiment

J. Geiss, P. Eberhardt, P. Signer, F. Buehler, and J. Meister

For several years, the fact has been established that $^4He^{2+}$ ions are present in the solar wind and that the relative abundance of the ions is highly variable (refs. 8-1 to 8-3). Helium-to-hydrogen values from 0.01 to 0.25 have been observed (ref. 8-1), but the average helium-to-hydrogen ratio in the solar wind is approximately 0.04 to 0.05 (refs. 8-1 and 8-4). At least during stationary conditions, the bulk velocities of hydrogen and helium are normally the same to within a few percent (refs. 8-1 and 8-5). The presence of ^3He and oxygen in the solar wind has also been reported recently (ref. 8-6). However, because these ion species were observed only under very favorable conditions, no values can be given for the averages and variations of their abundance.

Plasma and magnetic field measurements of Explorer 35 have established that, to a good approximation, the Moon behaves like a passive obstacle to the solar wind, and no evidence for a bow shock has been observed (refs. 8-7 and 8-8). Thus, during the normal lunar day, the solar-wind particles can be expected to reach the surface of the Moon with essentially unchanged energies. The grains of the fine lunar surface material contain great amounts of these particles. Consequently, it should be possible to extract valuable information on the composition of the solar wind from analyses of lunar surface material. However, the dust on the lunar surface is a solar-wind collector of uncertain properties, integrating the flux over an unknown period of time during which relative element abundances are probably changed significantly by both diffusion losses and saturation effects. No information on short-time variations can be derived from solar-wind particles implanted in lunar surface material; thus, the possibility that hydromagnetic processes influence the solar-wind composition cannot be assessed from data obtained only from lunar surface-material analysis.

For the foregoing reasons, it appeared worthwhile to carry out an experiment in which the solar wind would be sampled over a definite period of time by collecting solar-wind particles in a foil with well-defined trapping properties (refs. 8-9 and 8-10). The Apollo Solar-Wind-Composition (SWC) experiment was the first attempt at such solar-wind collecting.

Principle of the Experiment

An aluminum foil sheet 30 cm wide and 140 cm long with an area of approximately 4000 cm^2 was exposed to the solar wind at the lunar surface by the Apollo 11 crew. The foil was positioned perpendicular to the solar rays, exposed for 77 min, and brought back to Earth. Solar-wind particles, arriving with an energy of approximately 1 keV/nucleon, are expected to have penetrated approximately 10^{-5} cm into the foil, and a large fraction are expected to be firmly trapped. After release from the Lunar Receiving Laboratory (LRL), the foil will be analyzed for trapped solar-wind noble-gas atoms. Parts of the foil will be melted in ultrahigh vacuum systems, and the noble gases of solar-wind origin thus released will be analyzed with a mass spectrometer for element abundance and isotopic composition. If, during the exposure period, the solar wind has reached the foil with an intensity comparable to average flux values, then the Apollo 11 SWC experiment should allow an assessment of the abundances and isotopic compositions of helium, neon, and argon, despite the relatively short exposure time.

Instrumentation and Lunar Surface Operation

The hardware developed for the deployment of the foil carried on the lunar module (LM) descent stage Modularized Equipment Stowage Assembly is shown in figure 8-1. The hardware consists of a five-section telescopic pole and an aluminum foil screen rolled up on a reel. In the outbound configuration, the reel is stored inside the collapsed pole. The total weight is 0.95 lb. For deployment, the telescopic pole is extended, and the five sections lock automatically. The reel is then pulled out, and the foil is unrolled and fastened to a hook near the lower end of the pole. The reel is spring loaded to facilitate rewinding of the foil on the reel. The pole is pressed upright into the ground. Figure 8-2 shows the instrument as deployed by Astronaut Edwin E. Aldrin, Jr., at the lunar surface during the Apollo 11 extravehicular activity period. When figure 8-2 was photographed, the Sun was approximately 13° above the lunar horizon. One hr and 17 min later, Astronaut Neil A. Armstrong rolled up the foil on the reel. The foil (weight, 0.28 lb) was then detached from the telescopic pole, placed into a Teflon bag, and carried back to Earth inside the Apollo lunar sample return container, which also contained the documented sample.

FIGURE 8-2. — Deployed SWC unit with Astronaut Aldrin.

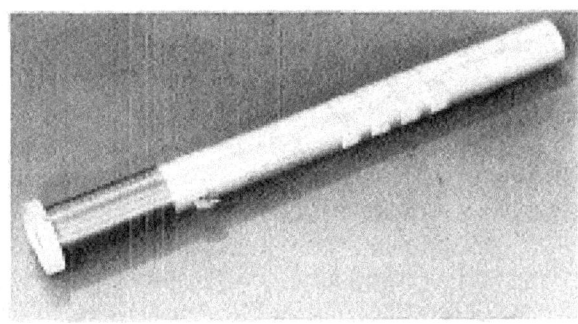

FIGURE 8-1. — Solar-Wind Composition experiment in stowed configuration ready for installation in the LM.

To eliminate the possibility of contamination of the lunar rocks in the container by the SWC instrument, special precautions were taken. All materials used in the construction of the reel and foil assemblies were analyzed for 13 geochemically important trace elements (lithium, beryllium, boron, magnesium, potassium, rubidium, strontium, yttrium, lanthanum, ytterbium, lead, thorium, and uranium). All concentrations were low enough to be geochemically acceptable.

To avoid organic contamination of the lunar surface material inside the container, the instrument was subjected to several heating cycles, and the degree of decontamination was controlled by mass spectrometric analysis. Prior to shipment to the NASA Manned Spacecraft Center and the NASA Kennedy Space Center for installation into the LM, the instrument was double bagged in Teflon and heat sterilized.

The Foil and Its Solar-Wind Trapping Properties

Dimensions and details of the exposed foil assembly are shown in figure 8-3. The main part of the assembly is a 15-μm-thick aluminum foil. The backside of the foil was anodically covered with approximately 1 μm of Al_2O_3, to keep the foil temperature below 100° C during exposure on the Moon. For reinforcement, the foil was rimmed with Teflon tape.

Test pieces 1, 3, and 5 are foils that were bombarded in the laboratory before the mission with a known flux of neon ions with an energy of 15 keV. The amounts of neon used were large by comparison with the expected solar-wind neon fluxes. Determination of the amount of neon implanted in these test pieces will indicate tensively investigated in the laboratory. Trapping probabilities for helium, neon, and argon were determined in a wide energy range. For average solar-wind energies, the probabilities are 89 percent for helium (3 keV), 97 percent for neon (15 keV), and 100 percent for argon (30 keV). Because these figures are only slightly energy dependent, they can be used even if, during exposure, the solar-wind velocity was quite different from the average. The trapping probabilities were found to be independent of foil temperature (20° to 120° C), and they were not affected by simultaneous bombardment with kiloelectron-volt hydrogen ions (H_2^+).

First Laboratory Investigations

The SWC return assembly was kept behind the primary biological barrier in the LRL until the quarantine was lifted, but a 1-ft² portion was cut from the midsection of the foil and sterilized inside a vacuum container by 125° C heat for 39 hr. The vacuum container, with this portion of the foil, was received at Berne, Switzerland, on August 12, 1969. Mass spectrometric investigation of the gas content in the container revealed little air leakage as indicated from the low argon content (3×10^{-8} cc STP ^{40}Ar). However, 10^{-6} cc STP helium and 10^{-8} cc STP neon were found. These gases are interpreted as having leaked, during sterilization, from lunar dust attached to the foil.

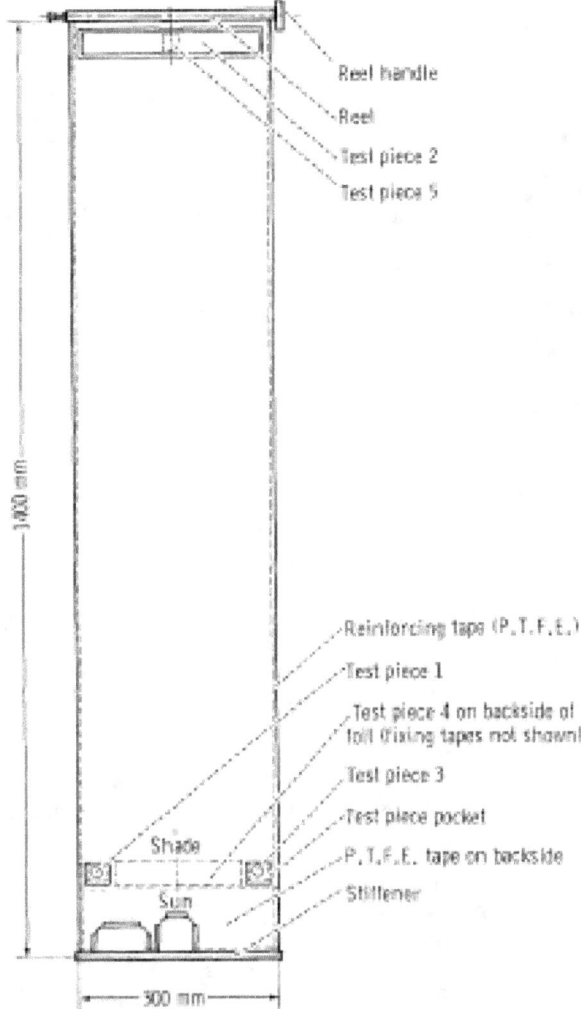

FIGURE 8-3.— Dimensions and details of the exposed foil assembly.

if solar-wind noble-gas losses could have occurred during the mission because of some unexpected diffusion or surface erosion processes. Test piece 5 was mounted in a closed pocket and remained shielded during the exposure on the lunar surface; test pieces 1 and 3 were exposed to the solar wind. Test piece 4 was a piece of foil taped to the backside of the main foil. A solar-wind flux coming from a direction opposite to the Sun could be distinguished by means of test piece 4. Test piece 2 was mounted in a position that remained shielded from the solar wind.

Before flight, the trapping properties of the foil material for kiloelectron-volt ions were ex-

With the unaided eye, virtually no dust particles were visible on the foil; however, microscopic investigation disclosed an appreciable number of fine particles. Figure 8–4 shows such dust grains, on the front side of the foil, photographed with a scanning electron microscope.

A few pieces of the foil, with areas of approximately 10 cm² each, were subjected to a cleaning process that included ultrasonic treatments. The noble gases were then extracted and analyzed in mass spectrometers. The presence of helium, neon, and argon was found, and the isotopic composition of these elements was measured. The element abundances and the isotopic compositions were clearly nonterrestrial and generally corresponded to the abundance estimates in the Sun (ref. 8–11) and to estimates derived from measurements taken in gas-rich

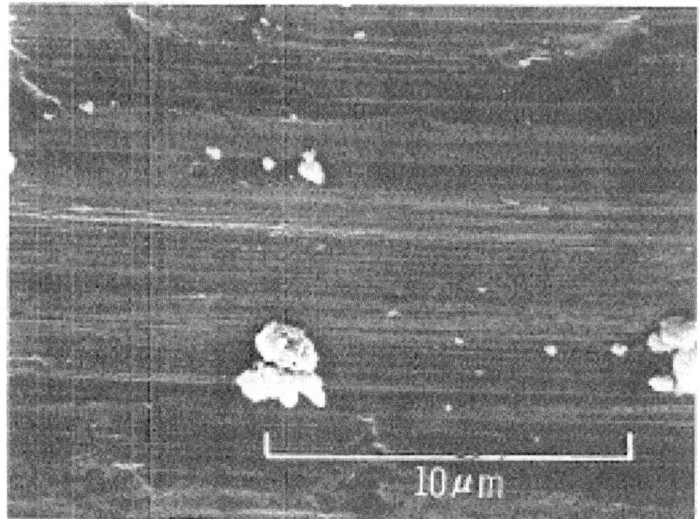

FIGURE 8-4. — Scanning electron microscope picture of front side of foil showing dust grains (4800 ×).

meteorites. At the time this report was written, September 8, 1969, various procedures calculated to decrease the dust level on the foil were in process, with the aim of obtaining an unequivocal distinction between solar-wind particles trapped in the foil and noble gases contained in the fine-grained lunar dust particles adhering to the foil.

References

8-1. HUNDHAUSEN, A. J.; ASBRIDGE, J. R.; BAME, S. J.; GILBERT, H. E.; and STRONG, I. B.: Vela 3 Satellite Observations of Solar Wind Ions: A Preliminary Report. J. Geophys. Res., vol. 72, no. 1, Jan. 1, 1967, pp. 87-99.

8-2. SNYDER, C. W.; and NEUGEBAUER, M.: Interplanetary Solar-Wind Measurements by Mariner 2. Proceedings of the Fourth International Space Science Symposium. Vol. IV, P. Muller, ed., North-Holland (Amsterdam), 1964, p. 89-113.

8-3. WOLFE, J. H.; SILVA, R. W.; MCKIBBIN, D. D.; and MATSON R. H.: The Compositional, Anisotropic, and Nonradial Flow Characteristics of the Solar Wind. J. Geophys. Res., vol. 1, no. 23, July 1, 1966, pp. 3329-3335.

8-4. SNYDER, CONWAY W.; and NEUGEBAUER, MARCIA: The Relation of Mariner 2 Plasma Data to Solar Phenomena. The Solar Wind, R. J. Mackin and Marcia Neugebauer, eds., Pergamon Press, 1966, p. 3.

8-5. OGILVIE, K. W.; BURLAGA, L. F.; and WILKERSON, T. D.: Plasma Observations on Explorer 34. J. Geophys. Res., vol. 73, no. 21, Nov. 1, 1968, pp. 6809-6824.

8-6. BAME, S. J.; HUNDHAUSEN, A. J.; ASBRIDGE, J. R.; and STRONG, I. B.: Solar Wind Ion Composition. Phys. Rev. Letters, vol. 20, no. 8, Feb. 19, 1968, pp. 393-395.

8-7. LYON, E. F.; BRIDGE, H. S.; and BINSACK, J. H.: Explorer 35 Plasma Measurements in the Vicinity of the Moon. J. Geophys. Res., vol. 72, no. 23, Dec. 1, 1967, pp. 6113-6117.

8-8. NESS, N. F.; BEHANNON, K. W.; SCEARCE, C. S.; and CANTARANO, S. C.: Early Results from the Magnetic Field Experiment on Lunar Explorer 35. J. Geophys. Res., vol. 72, no. 23, Dec. 1, 1967, pp. 5769-5778.

8-9. SIGNER, PETER; EBERHARDT, PETER; and GEISS, JOHANNES: Possible Determination of the Solar Wind Composition. J. Geophys. Res., vol. 70, no. 9, May 1, 1965, pp. 2243-2248.

8-10. BUEHLER, F.; GEISS, J.; MEISTER, J.; EBERHARDT, P.; HUNEKE, J. C.; and SIGNER, P.: Trapping of the Solar Wind in Solids. Earth Planet. Sci. Lett., vol. 1, 1966, pp. 249-255.

8-11. ALLER, LAWRENCE HUGH: The Abundance of the Elements. Interscience Publishers, 1961.

9. Lunar Surface Closeup Stereoscopic Photography

The lunar samples returned by the Apollo 11 mission have provided preliminary information about the physical and chemical properties of the Moon and, in particular, about Tranquillity Base. Because of the mechanical environment to which lunar samples are subjected during their return from the Moon, limited information can be obtained from lunar samples about the structure and texture of the loose, fine-grained material that composes the upper surface of the lunar crust. A stereoscopic camera capable of photographing the small-scale (between micro and macro) lunar surface features was suggested by Thomas Gold, Cornell University, and built under contract for NASA.

The photographs taken on the mission with the closeup stereoscopic camera are of outstanding quality and show in detail the nature of the lunar surface material. Several photographs contain unusual features. From the photographs, information can be derived about the small-scale lunar surface geologic features and about processes occurring on the surface. This chapter presents a description of the closeup stereoscopic camera, lists and shows single photos from pairs available for stereoscopic study, and contains an interpretation of the results reported by Professor Gold in *Science*, vol. 165, no. 3900, pp. 1345-1349, Sept. 26, 1969.

Apollo Lunar Surface Closeup Camera

The Apollo Lunar Surface Closeup Camera (ALSCC), built by Eastman Kodak under contract to the NASA Manned Spacecraft Center, was designed to optimize operational simplicity (fig. 9-1). The camera is an automatic, self-powered, twin-lens stereoscopic camera capable of resolving objects as small as 80 μm in diameter on color film. Simplicity of operation was achieved by making the camera focus and the flash exposure fixed and by loading the camera prior to launch with sufficient Ektachrome MS (S0368) film for the complete mission. To obtain a photograph, an astronaut merely sets the camera over the material to be photographed and depresses the trigger located on the camera handle. When the exposure is complete, the film is automatically advanced to the next frame, and the electronic flash is recharged.

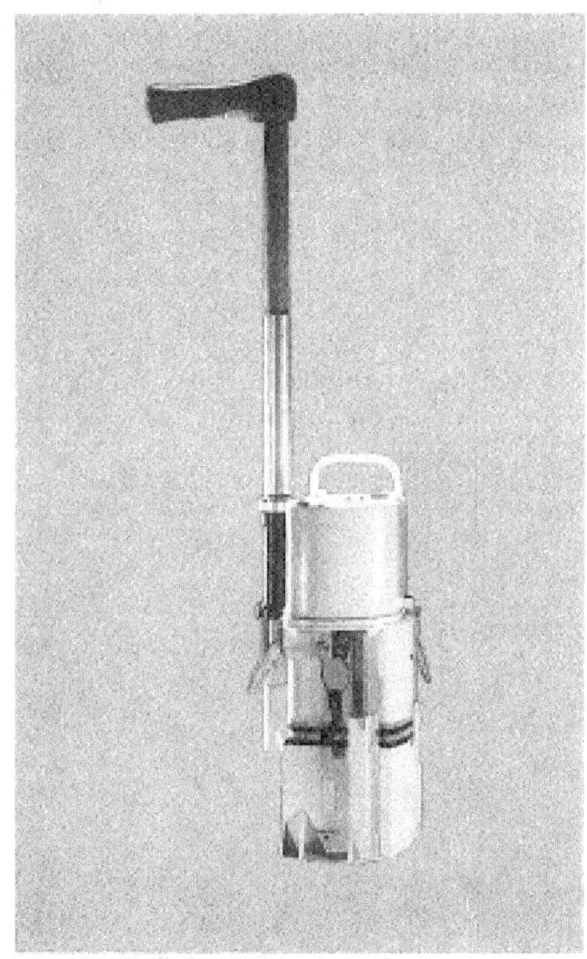

FIGURE 9-1. — Apollo lunar surface closeup camera.

The requirement to simplify the ALSCC operation necessitated a reasonable depth of field that could be achieved only by limiting the image magnification produced by the system. The camera lenses are diffraction-limited 46.12 mm f/17 Kodak M-39 copy lenses focused for an object distance of 184.5 mm, providing an image magnification of 0.33. The lenses are mounted 29 mm apart, with their optical axes parallel. The area photographed is 72 mm x 82.8 mm and centered between the optical axes; thus, stereoscopic photographs with a base-to-height ratio of 0.16 are provided (fig. 9-2).

Location of the Stereoscopic Photographs

The specific location at the Apollo 11 lunar landing site where individual stereoscopic photographs were taken cannot be determined with a high degree of confidence. Because of the limited size of the area photographed by the camera, the subject material cannot be identified within the large-scale lunar surface photography. In the scientific debriefing of the Apollo 11 crew, Astronaut Armstrong indicated the general areas, relative to the lunar module, where sequential sets of photographs were taken. Based upon Astronaut Armstrong's comments and the time history of events on the surface, the general locations of the photographs were tentatively established (fig. 9-3).

Description of Photographs

Analysis of the photographic data is not complete and will therefore be published in future documents. For this report, only a preliminary description of four photographs, provided by Gold, is included.

A stereoscopic view (fig. 9-4) shows a close-up of a small lump of lunar surface powder approximately 0.5 in. across, with a splash of glassy material over it. A drop of molten material appears to have fallen on the powder, splashed, and frozen.

Figure 9-5 shows a clump of lunar surface powder with various small, differently colored pieces. Many small, shiny spherical particles can be seen.

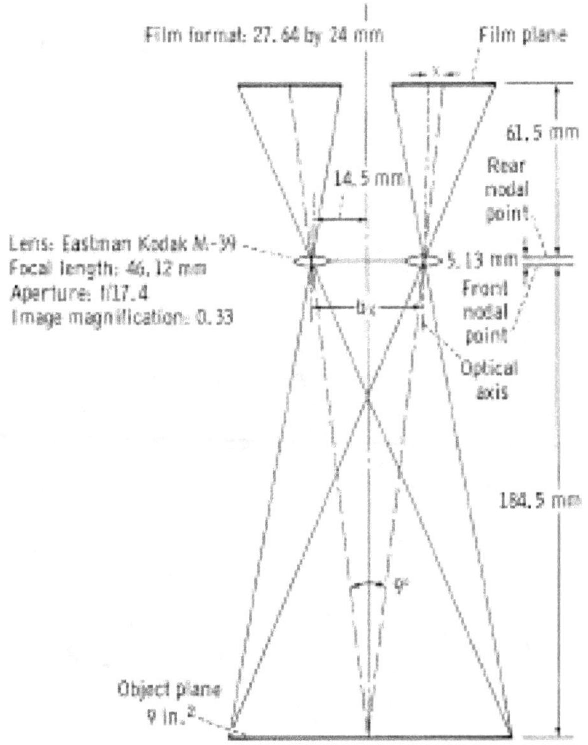

FIGURE 9-2. — Schematic of the ALSCC.

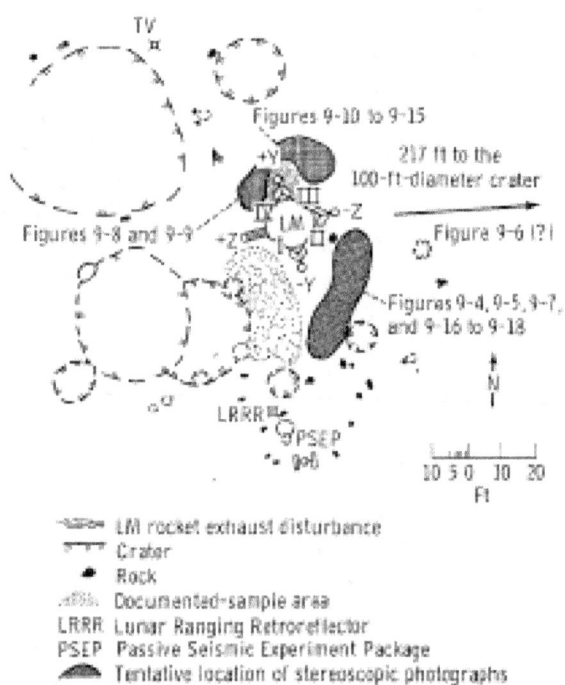

FIGURE 9-3. — Locations of ALSCC photographs.

FIGURE 9-4. — A small lump of lunar soil with a glazed top and a smaller glazed piece to the right. The larger object shows no change of color or texture compared with the surrounding ground, known to be soft, fine-grained soil, except for the glaze mark. (NASA AS11-45-6704)

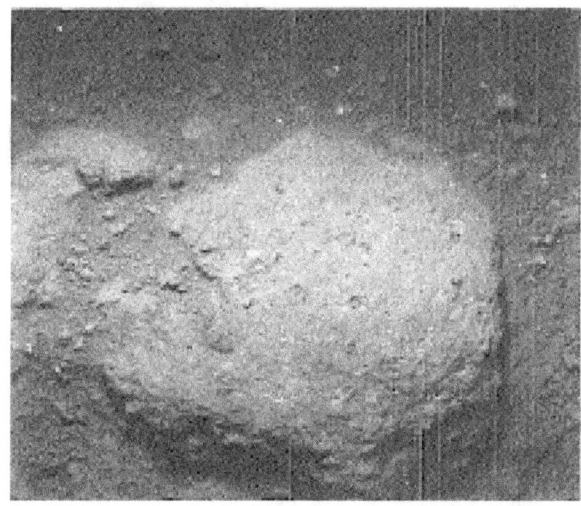

FIGURE 9-6. — Stone embedded in powdery lunar surface material. (NASA AS11-45-6712)

Another stereoscopic view (fig. 9–6) shows a stone, approximately 2.5 in. long, embedded in the powdery lunar surface material. The small pieces grouped closely around the stone suggest that it suffered some erosion. On the surface, several small pits are seen, mostly less than one-eighth in. in size and with a glazed surface. The pits have a raised rim, characteristic of pits made by high-velocity micrometeorite impacts.

One stereoscopic view (fig. 9–7) of the surface of a lunar rock shows an embedded ¾-in. fragment of a different color. On the surface, several small pits are seen, mostly less than one-eighth in. in size and with a glazed surface. The pits have a raised rim, characteristic of pits made by high-velocity micrometeorite impacts.

FIGURE 9-5. — Clump of lunar surface powder (NASA AS11-45-6706)

FIGURE 9-7. — Stereoscopic view (NASA AS11-45-6709)

The stereoscopic views in figures 9-8 to 9-20 are included in this report to make the availability of the photographs known. Descriptions or interpretation of these photographs were not available for this publication.

These photographs are available upon request from the NASA Manned Spacecraft Center, Houston, Tex. The following is a listing of the 17 photograph numbers, keyed to the figure number used in this chapter.

Figure	Photograph
9-4	AS11-45-6704
9-5	AS11-45-6706
9-6	AS11-45-6712
9-7	AS11-45-6709
9-8	AS11-45-6697
9-9	AS11-45-6698
9-10	AS11-45-6699
9-11	AS11-45-6700
9-12	AS11-45-6701
9-13	AS11-45-6702
9-14	AS11-45-6702-1
9-15	AS11-45-6703
9-16	AS11-45-6705
9-17	AS11-45-6707
9-18	AS11-45-6708
9-19	AS11-45-6710
9-20	AS11-45-6713

FIGURE 9-9. — Stereoscopic view (NASA AS11-45-6698)

FIGURE 9-8. — Stereoscopic view (NASA AS11-45-6697)

FIGURE 9-10. — Stereoscopic view (NASA AS11-45-6699)

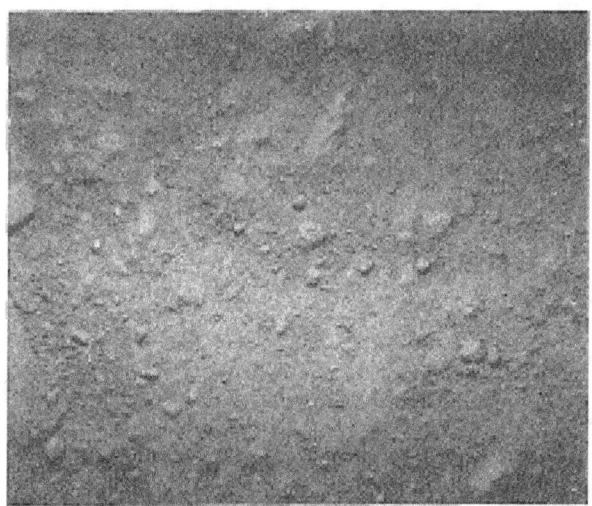

FIGURE 9-11. — This is the bottom of a crater, and the affected area is close to the upper edge of the frame (and presumably beyond). Here, glazing is seen in a very regular fashion, with the edges and points that face upwards and out in a direction to the upper right consistently affected. The stereoscopic view is required to observe the directional effect. (NASA AS11-45-6700)

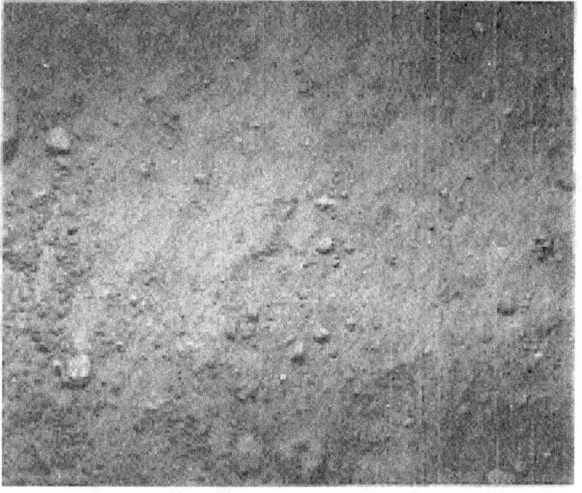

FIGURE 9-13. — Stereoscopic view (NASA AS11-45-6702)

FIGURE 9-12. — Stereoscopic view (NASA AS11-45-6701)

FIGURE 9-14. — Stereoscopic view (NASA AS11-45-6702-1)

FIGURE 9-15. — Stereoscopic view (NASA AS11-45-6703)

FIGURE 9-17. — Stereoscopic view (NASA AS11-45-6707)

FIGURE 9-16. — Stereoscopic view (NASA AS11-45-6705)

FIGURE 9-18. — A large area of glazing is seen in the lower left region, both on the horizontal and on the sloping surfaces of the protuberance. Some glazing is also seen on the lower right object, with edges and points particularly affected. In the upper part of the picture a small glossy bead is seen poised on a narrow pedestal. (The detail in the color stereoscopic transparencies far exceeds that which can be reproduced in the present print.) (NASA AS11-45-6708)

FIGURE 9-19. — Stereoscopic view (NASA AS11-45-6710)

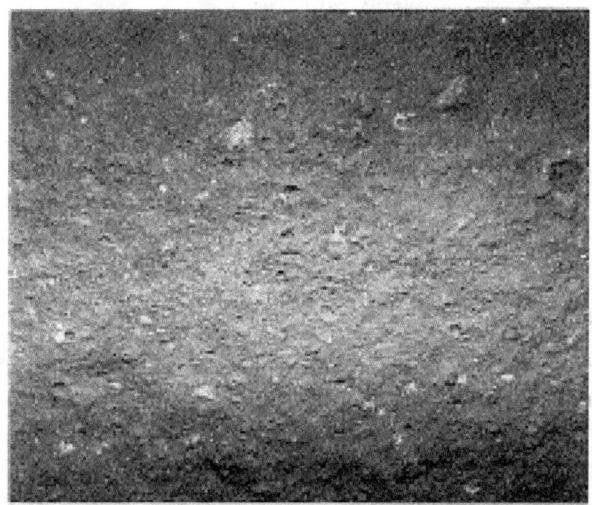

FIGURE 9-20. — Stereoscopic view (NASA AS11-45-6713)

Apollo 11 Observations of a Remarkable Glazing Phenomenon on the Lunar Surface

T. Gold

The Apollo 11 mission carried a closeup stereo camera with which the astronauts took 17 pictures. Each is of an area 3 by 3 in., seen with a resolution of approximately 80 μm. There are many details seen in these pictures that were not known previously or that could not be seen with similar definition by Astronauts Armstrong and Aldrin in their careful inspection of the lunar surface. One of those is outstanding in being wholly unexpected and possibly of consequence not only to lunar geology but also to aspects of the study of the Sun, the Earth, and other bodies of the solar system.

The observation is that of glossy surfaces, in appearance a glass of color similar to the surrounding powdery medium, lying in very particular positions. Small craters that are plentiful on the lunar surface — 6 in. to 2 or 3 feet in diameter — have frequently some clumps of the lunar soil or rough spots of the surface texture concentrated toward their center. They give the appearance of having been swept in at a time later than the formation of the crater. Some of these little lumps appear to have just their top surfaces glazed. The glassy patches that can be seen on the photographs range in size from ½ mm to about 1 cm.

The glazed areas are clearly concentrated toward the top surfaces of protuberances, although they exist also on some sides. Points and edges appear to be strongly favored for the glazing process. In some cases, droplets appear to have run down an inclined surface for a few millimeters and congealed there.

The astronauts' information was that similar things were seen in many small craters, many more than the eight instances of which we have photographs. They were not able at the time to pick up any of the objects for the sample return. They did not see a single glazed piece in any different geometrical position other than near the center of a small crater. They were not able to handle any of the objects, and thus have no direct observation as to the thickness of the glass. Since they only saw them from eye-height, looking down at the ground in front of them, their visual acuity would be considerably less than that of the camera, and they could therefore not supplement the detailed information contained in the pictures.

Several theories of the origin of this phenomenon have to be discussed.

1. An effect connected with the rocket exhaust of the descent stage.

It might be thought that in the last phases of the landing process the rocket flame melted some lunar material and that this was blasted over the surrounding terrain. While at first this seems a very attractive hypothesis, when pursued in more detail it runs into difficulties. These are concerned with the regional distribution of the objects and also with the detailed geometrical relationships in the places where they are seen.

The astronauts noticed these objects first in a region ahead and to one side of the landing approach path, about 50 ft from the spacecraft. The pattern of disturbance caused by the rocket could be seen to extend hardly beyond the footpads of the spacecraft. The inspection of the approach path and the region directly underneath the nozzle of the rocket showed the ground disturbed there, but no evidence of any melting was seen. It seems most unlikely that if melting occurred, all molten material was blasted away, not leaving a higher concentration near the source.

No explanation can be found of how material so flung out would end up concentrated in little clumpings in the bottoms of small craters without there being also some on the flat surface or in much shallower depressions. The material could not be thought to have been propelled by gas drag and to have slid over the surface, to land perhaps in sheltered holes, since the rocket gas-drag forces are quite inadequate at a distance of more than a few feet from the rocket. This is substantiated by the fact that the radially streaked pattern of the ground radiating out from underneath the spacecraft disappears at a distance of between 10 and 15 ft. No glassy objects could have been dragged over the surface by this stream at a distance of 50 ft; if they had been flung that far, they would be lying essentially where they landed, and no reason is then seen for the specific distribution that was found.

The glassy objects were, by the astronauts' descriptions, substantiated by the pictures, usually found in a tight clumping, and the pictures indicate in several cases a definite common azimuth dependence of the effect among the members of any one group. (There is no record of the azimuth position of the camera, so different groups cannot be compared with each other; but the camera was generally held facing nearly vertically downward. The astronauts were not aware of any azimuth effect, nor could they have been expected to see it, since it becomes evident only from a careful study of the high-resolution pictures.)

It thus seems extremely improbable that any effect of the rocket exhaust can be held responsible for the phenomenon.

2. Splashing of liquid drops from a larger impact elsewhere.

Many of the same objections can be raised again here. Such droplets could not have landed in preferential places. On impact, with the velocities necessary from a source more than a few feet away, they would have toppled and destroyed many of the frail structures that they hit, and thus the geometrical relationships would have been lost.

3. The objects are common in the ground, where they were distributed after being created by shock heating or volcanism. Erosion has exposed some of them.

The fact that the objects are found in the bottoms of little craters argues against slow erosion exposing them, for these craters are filled up and not emptied out by the lunar surface transportation processes. There are many cases of a clearly visible outline of the glazing, showing shapes characteristic of viscous flow defining that outline. One is clearly not dealing with entirely glazed objects of which only a certain fraction is cleaned off and exposed, but rather one is dealing with objects that are covered with a glaze mainly on top.

Small glassy spheres had been reported to be common in the lunar soil, and many are indeed also seen on the pictures. Their origin may be understood merely as a consequence of meteoritic impact melting and freezing during the ballistic trajectory of the ejected material. If the ground has been plowed over many times by impacts, as suggested by the topography of the Moon, a substantial admixture of spheres and other shapes resulting from only partial melting must indeed be expected in the ground. The glazed surfaces discussed here appear not to be of this character. Fragments of them may also appear in the lunar soil, and those would be expected to be thin pieces, shiny on one side, and rough with embedded soil on the other.

4. They are objects created by the impact that made the little craters in which they are now found.

This seems impossible in view of the fact that they are frequently quite frail structures, apparently only made of the powdery material, and they could not have survived and remained there in definite geometrical positions when an explosive force excavated the crater. While in a hard rock a glazed coating may remain lining the impact pit, this is very unlikely in a soft soil. All material exposed to enough heat to melt would be exposed also to aerodynamic forces large enough to be expelled. Any glass made by a strong pressure wave in the ground and subsequently exposed would not have glossy surfaces except as a result of fracture. The shapes seen are in most cases clearly the result of a surface viscous flow and not of a fracture.

5. Radiation heating.

An intense source of radiative heating could account for all the facts that are now known. The wavelength of this radiation could be anywhere in the electromagnetic spectrum between the far ultraviolet and the far infrared. For radiative heating from a small angular source in the sky the bottoms of craters are substantially favored. The equilibrium temperature in a given radiation field is likely to be between 10 and 20 percent in absolute temperature above that of flat ground. (This problem has been discussed in some detail by Buhl, Welch, and Rea (ref. 9–1).) The reason for that is, of course, the diminution of the solid angle subtended by the cold sky for a particle in a hollow compared with one on flat ground. There is thus an intensity range where radiation heating will cause melting only in the bottoms of craters. The positioning of the objects can be understood in these terms.

No significant mechanical forces would be involved, and frail structures could suffer melting of their upper surface without being deranged. One protuberance, for example, shows no change in color or texture from the surrounding medium known to be soft powder; yet it has a large patch of glazed material over the top of it. If the source of the heating were not directly overhead, there could be a common azimuthal preference for the melted surfaces.

Protuberances, rather than flat ground, and sharp points and edges would be favored for melting if the heat source existed for a short time only. The reason for this is that heat conduction will carry less heat away from such positions, and they would reach the temperature corresponding to the local radiative equilibrium sooner than would flat surfaces. The observations are very clear in demonstrating such an effect. Similarly, the apparently thin coating of glaze clearly seen in many of the pictures indicates a short duration for the period of the radiative heating. With the approximately known rate of heat transfer in the lunar material, one may estimate that the duration of the heating phase was between 10 and 100 sec.

Among the interpretations discussed here, only radiative heating seems to be able to account for the major observational data; and it accounts well for both the distribution over the ground and the detailed geometrical arrangements in which the glazing is found, for it is in just those places that the temperature would be expected to be at a maximum. We must therefore seek clues as to the nature of such a flash of radiation.

The time of occurrence of the flash heating would have to have been sufficiently recent for micrometeorites not to have destroyed the glaze, nor for the mechanisms that redistribute the lunar soil to have blanketed the objects. Estimates of the micrometeorite rate at the present time vary considerably, but it seems very unlikely that the glaze could be maintained on the surface for as much as 100 000 yr and probably not for more than 30 000. The event in question would thus have to have taken place within this geologically very recent past. Several possibilities have to be discussed.

a. An impact fireball on the Moon.

A large impact would no doubt generate intense radiation from the explosive gas cloud. This radiation will, however, be largely screened from reaching areas outside the crater made by the impact, and it seems unlikely that high radiation intensities would extend over regions beyond those swept over by the explosive action or the expelled material. Craters would be particularly unfavorable locations for such a source of heat, rather than highly favored as the observations show.

b. An impact fireball on the Earth.

Although very large impacts may have taken place on the Earth within geologic time, and may indeed be responsible for some of the widespread falls of tektites, no really major event is thought to have occurred within the sufficiently recent past. The energy released in an impact on the Earth of a 1-km meteorite at a velocity of 20 km/sec is of the order of 3×10^{27} ergs and that, if it were all converted into radiation, would deposit on the order of 4×10^{8} ergs/cm^2 on the lunar surface. The value required to cause the observed effect is of the order of 3×10^{9} ergs/cm^2. No fireball of adequate intensity seems to have existed on the Earth within the last hundred thousand years.

c. A flash from the Sun.

One can calculate the increase in the solar emission that would be required to cause melting in favorable places on the lunar surface. The normal noontime temperature on the lunar equator is known to be approximately 394° K. The analysis of Buhl et al. (ref. 9–1) shows that the absolute temperature in the bottom of a hemispherical crater must be expected to rise about 11 percent above that of the surrounds. The normal noontime temperature would thus be 437° K in such locations. If f is the factor by which the solar luminosity is increased, we require that

$$f = \left(\frac{T_m}{T_0}\right)^4$$

where T_m is the melting temperature and T_0 the normal temperature in the same location. If we adopt a melting temperature of the lunar surface material of 1400° K, an appropriate figure for basalt, the value of f would be 106. The temperature of the flat ground in the same radiation field would then be 1260° K. (For a brief flash the fact that a crater bottom started out 40° hotter would further increase the margin between melting there and elsewhere.)

The flash heating, if it originated from the Sun, overhead at the time in the lunar region in question, thus requires that the Sun's luminosity flared up for a period of between 10 and 100 sec to a luminosity of more than 100 times its present value. What could be the origin of such a flareup?

Solar flares, as they are known at the present time, only reach a total energy output of less than 10^{23} ergs. The quantity we are discussing here is at least 4×10^{36} ergs for an isotropic source, or 2×10^{36} ergs for a localized source on the Sun that radiates into a hemisphere only. While it cannot be ruled out completely that flares of such an intensity could occur, we have no evidence in this regard, although some earlier speculations exist (ref. 9–2).

The phenomenon may not be in the nature of a flare, but in the nature of a very minor novalike outburst of the Sun. The nova phenomenon among the stars is not understood, but it is normally concerned with a different class of star, and the intensity of the outburst, up to 100 times the normal luminosity, for a period of only tens of seconds every few tens of thousands of years, would be a phenomenon that would not have been recognized astronomically among stars of the solar type. One can therefore not completely rule out the possibility that stars like the Sun do have occasional instabilities of internal origin, but very much weaker than the novae.

The infall into the Sun of a comet or asteroid is another possibility.[1] An object falling into the Sun at the velocity of escape would have to have a mass of 3×10^{21} g for the kinetic energy to be sufficient to generate the flash. This would put the object into the class of asteroids or very large cometary cores.

For the infalling energy to be converted into radiation in its entirety, it would be necessary for the object to break up in the solar atmosphere, and for no substantial pieces to remain intact at the level of the photosphere. Tidal disruption and violent heating could perhaps be invoked to break up a cometary body more readily than an asteroidal one. Also, comets have been known to approach the Sun very closely in historical times, and the statistical evidence would be that many have hit the Sun in a period of 30 000 years. The mass required (3×10^{21} g) is above that which has been guessed at for presently known comets.

[1] Suggested by F. Hoyle.

An icy cometary core would have to have a radius of approximately 100 km, while some presently known comets[2] may have cores as large as 50 km. The suggestion that the flash was due to an infalling comet seems to merit further attention.

If such a flash occurred on the Sun within geologically recent times, there are many other consequences that may perhaps be recognized on the Earth or other planets. If the flash occurred as a result of an infalling object, or of a superflare, or of an internal outburst that exposed subphotospheric material, its spectrum would be mainly in the ultraviolet. So far as the Earth is concerned, the total extra heat delivered would be mainly spread in the atmosphere and does not cause a substantial rise in the temperature of the lower atmosphere. The heat delivered to the ground is not likely to have been enough to cause any permanent effects. Nevertheless, there may be effects in the geologic record that can be attributed to such an event. The upper atmosphere of the Earth and of Venus, and the entire atmosphere of Mars, would, however, have been seriously affected. On the Earth, the ratio of ^{3}He and ^{4}He in the outer atmosphere suggests that there is no substantial preferential loss by slow evaporation of ^{3}He (refs. 9–3 and 9–4). If the outer atmosphere were entirely swept away by such outbursts, then the present ratio, which implies an absence of a preferential thermal evaporation at normal temperatures of ^{3}He, would be more readily understood. The apparent absence of nitrogen on Mars, as reported from the data of the Mariner 6 and 7 flights, could also be understood if the entire atmosphere had been swept away and, in the short time since, had been replenished only from the constituents frozen out on the polar cap, which may be CO_2, but not nitrogen.[3]

[2] F. Whipple, private communication.
[3] Suggested by C. Sagan.

Flash heating on the surface of Mercury would have been much more intense, and if made of lunar-type rock, almost an entire hemisphere would have been glazed. We do not know the surface material, nor the effects of erosion and solar irradiation on Mercury, and the absence now of a glazed hemisphere could be understood as due to a different material or a faster destruction of glass on Mercury than on the Moon.

No doubt many more facts will come to light concerning the lunar surface and, perhaps, this phenomenon, in the analysis of the lunar samples brought back by Apollo 11. A rapid publication of these findings was nevertheless indicated by the urgency to make the best possible scientific preparations for the next Apollo flights.

A more complete review of this and other information contained in the closeup pictures will be published later.

References

9-1. BUHL, D.; WELCH, W. J.; and REA, D. G.: J. Geophys. Res., vol. 73, 1968, p. 5281.
9-2. GOLD, T.: In Semaine d'Etude sur le Probleme du Rayonnement Cosmique dans l'Espace Interplanetaire (Pontificiae Academiae Scientiarum Scripta Varia, Vatican), 1963, pp. 159–174.
9-3. MACDONALD, G. J. F.: Rev. Geophys., vol. 1, 1963, p. 305.
9-4. AXFORD, W. I.: J. Geophys. Res., vol. 73, 1968, p. 6855.

ACKNOWLEDGMENTS

I am indebted to the National Aeronautics and Space Administration for arranging for the construction and deployment of the closeup camera; to E. Purcell, E. Land, J. Baker, R. Scott, and F. Pearce for their contributions in outlining this camera project; to Eastman Kodak for the design and fabrication of the instrument; and chiefly to N. Armstrong for his successful use of it, and his excellent descriptions of all relevant lunar surface observations made by himself and E. Aldrin. Work on the lunar closeup camera and the analysis of photographs obtained with it is supported at Cornell under NASA Contract Number NAS9–9017. I am grateful to F. Hoyle, F. L. Whipple, G. J. F. MacDonald, and S. Soter for helpful discussions.

10. The Modified Dust Detector in the Early Apollo Scientific Experiments Package

J. R. Bates, S. C. Freden, and B. J. O'Brien

The modified dust detector (fig. 10-1) in the Early Apollo Scientific Experiments Package (EASEP) consists of three n-on-p silicon solar cells mounted on the top horizontal surface of the dust-detector housing. The objectives of the dust-detector experiment, which is also referred to as the Dust, Thermal, and Radiation Engineering Measurements Package (DTREM I), are to measure dust accumulation resulting from the LM ascent or from any long-term cause and to measure radiation degradation of the voltage output from the solar cells. The radiation degradation is caused by solar-flare particle events.

The concept of radiation measurement by a dust detector is fairly straightforward. Radiation degradation of the voltage output from a solar cell is predictable. If a glass shield is placed over the surface of a solar cell, particles above certain energies will damage the cell. The particle energy required to damage the cells is determined by the thickness of the glass shield.

A preliminary analysis of data received in the Mission Control Center (MCC) on the first lunar day that the experiment was operational showed no appreciable cell degradation caused by dust or debris from the LM ascent. Shadows from the antenna, the antenna alining pole, and the carrying handle (fig. 10-2) were evident on the dust-detector outputs from day 205.5 to day 208 (figs. 10-3 and 10-4).

In the preliminary analysis, the voltage outputs from the solar cells on the dust detector were 10 percent less at lunar noon than prelaunch predictions had indicated. These lower-than-predicted voltage outputs may have been caused by dust, but they may also have been a result of errors in the prelaunch calibration measurements of the cells and errors in the calibration data used in the MCC. Figures 10-3 and 10-4 are samples of the real-time data received in the MCC.

The maximum temperature of the solar cells on the dust detector during the first lunar day of the experiment was approximately 104° C. However, this temperature was recorded after lunar noon, and shadowing of the dust detector (from the antenna, the antenna alining pole, and the carrying handle) occurred just before lunar noon. Without shadowing, the temperature might have been 5° to 10° C higher.

The dust-detector data obtained during the sunset period of the lunar day appear to be

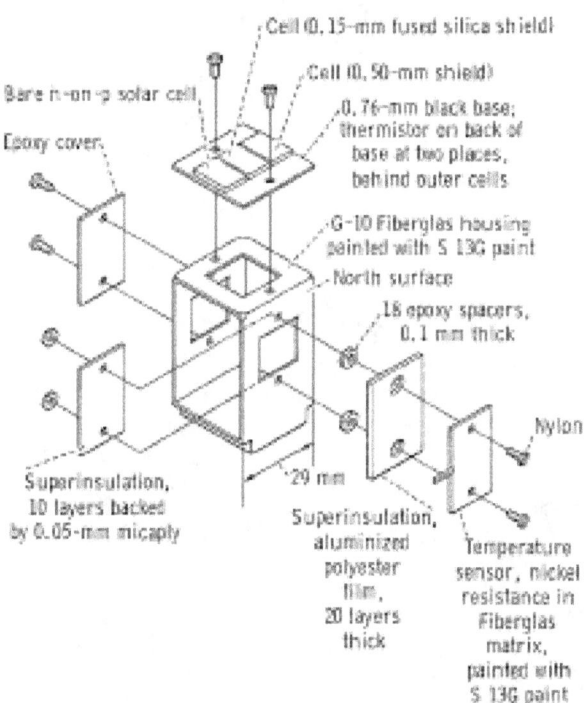

FIGURE 10-1. — Dust, Thermal, and Radiation Engineering Measurement Package (modified dust-detector experiment) on the Apollo 11 EASEP.

200   APOLLO 11 PRELIMINARY SCIENCE REPORT

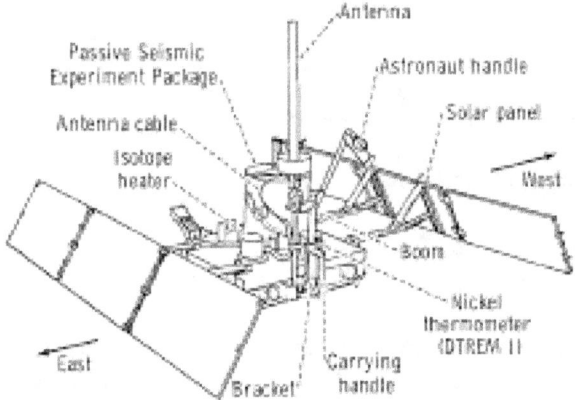

FIGURE 10-2. — Deployed configuration of the Passive Seismic Experiment Package showing the geometry of the DTREM I.

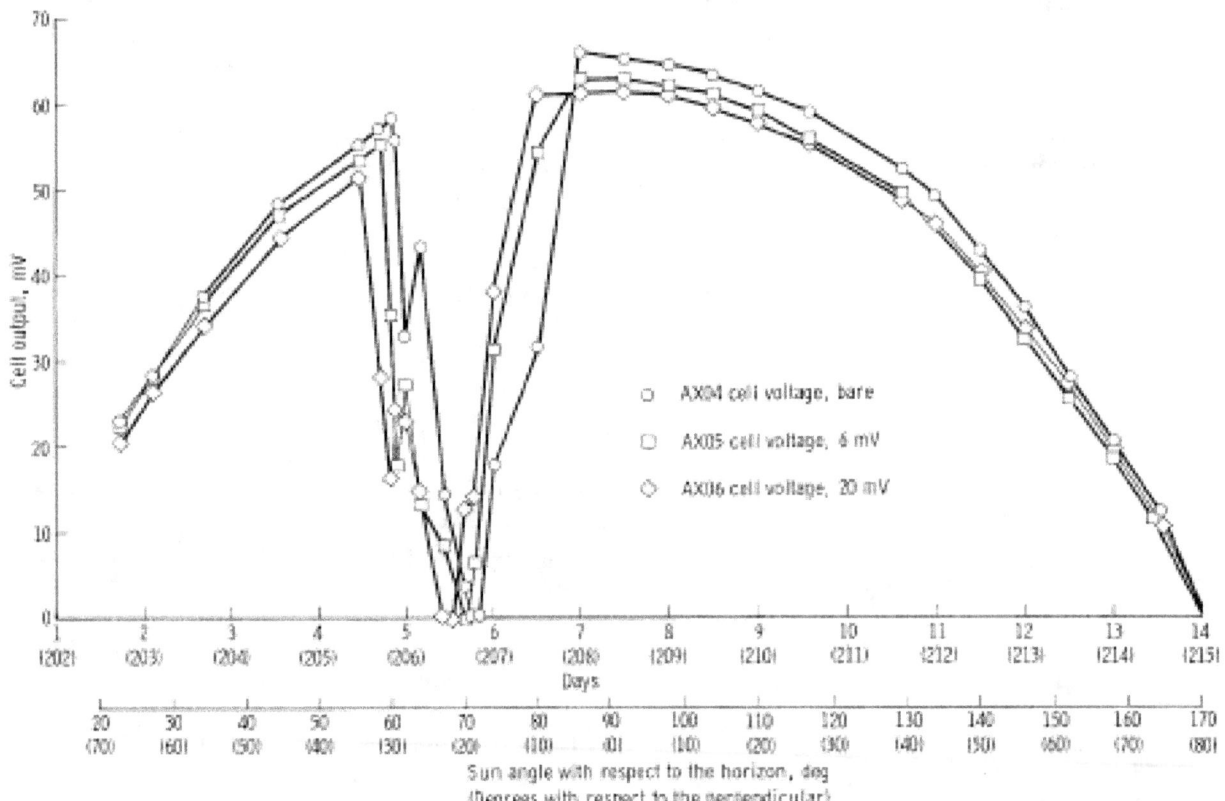

FIGURE 10-3. — Dust-detector solar-cell output.

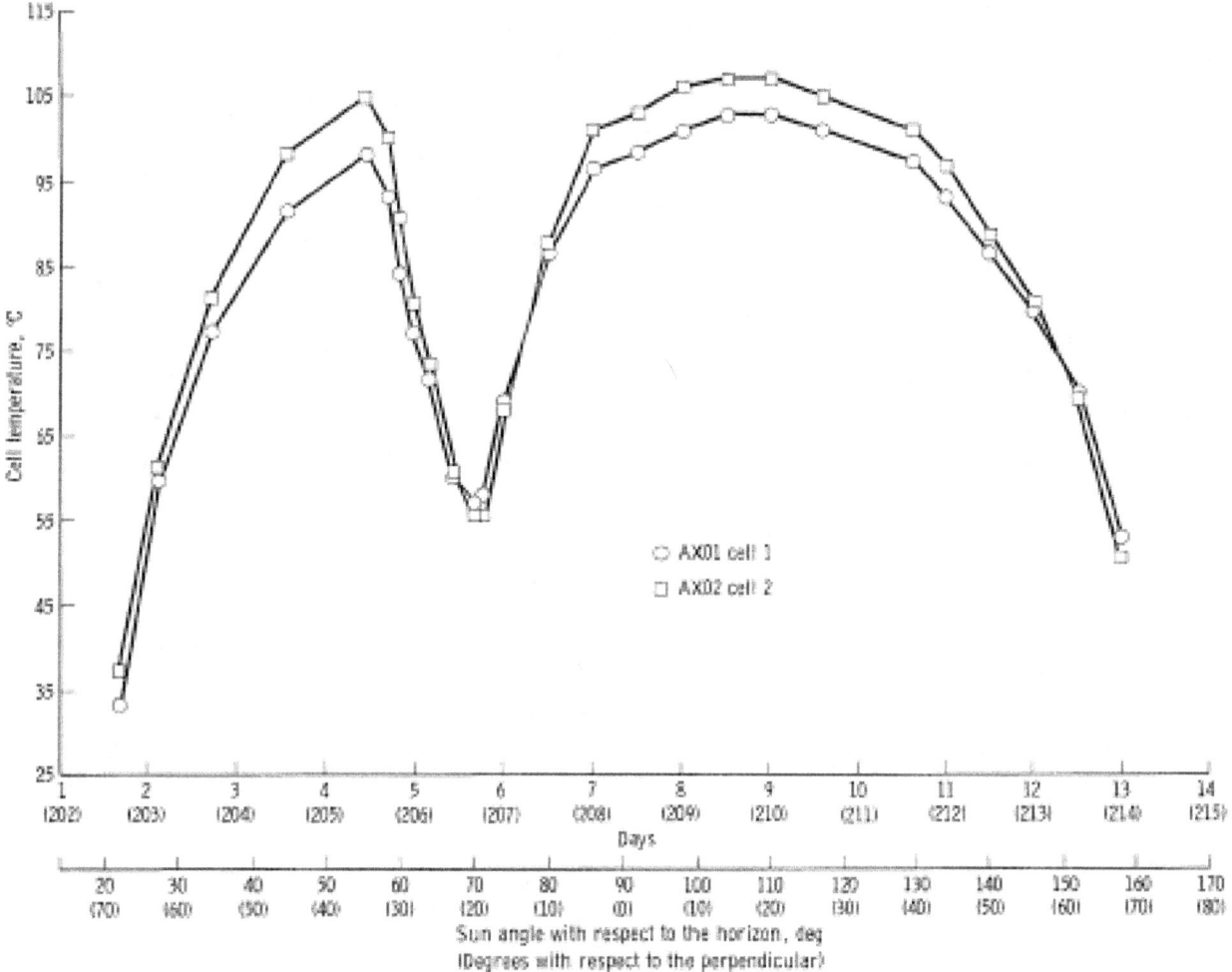

FIGURE 10-4. — Dust-detector solar-cell temperature.

slightly offset from similar data obtained during the sunrise period. This slight offsetting may have been caused by the dust detector being tilted 2° to 5° from the horizontal, by the nonzero declination of the Sun, or by both. Also, during sunset, the reflections and scattered light from the surrounding EASEP components are different from the reflections and scattered light during sunrise.

To detect degradation of the solar cells, data obtained on the first lunar day of the experiment will be compared with data obtained on successive lunar days. After sufficient data have been accumulated, a detailed analysis and establishment of baseline conditions will be made by using off-line computer routines that contain exact calibration data.

APPENDIX A
Glossary of Terms

ablation — removal; wearing away.

albedo — ratio of light reflected to light incident on a surface.

arcuate — in the form of a bow.

augite — a mineral consisting of an aluminous, usually black or dark-green, variety of pyroxene.

basalt — a dark-gray to black, dense to fine-grained igneous rock consisting of basic plagioclase, augite, and usually magnetic with olivine or basalt glass (or both) sometimes present.

bleb — small bit or particle of distinctive material.

botryoidal — having the form of a bunch of grapes.

breccia — a rock consisting of sharp fragments embedded in a fine-grained matrix.

clast — rock composed of fragmental material of specified types.

clino-pyroxene — a mineral occurring in monoclinic, short, thick, prismatic crystals, varying in color from white to dark green or black (rarely blue).

coudé focus — focus at a fixed place along the polar axis of a telescope.

cristobalite — silica occurring in white octahedra.

diabase — a fine-grained igneous rock of the composition gabbro, but with lath-shaped plagioclase crystals enclosed wholly or in part in later formed augite.

dike — tabular igneous rock that has been injected (while molten) into a fissure.

diktytaxitic — volcanic rock of clastic appearance.

equant — having crystals with equal or nearly equal diameters in all directions.

euhedral — refers to minerals whose crystals have had no interference in growth.

exfoliation — the process of breaking loose thin concentric shells or flakes from a rock surface.

feldspathic — pertaining to feldspar.

fines — very small particles in a mixture of various sizes.

gabbro — a granular igneous rock composed essentially of calcic plagioclase, a ferromagnesian mineral, and accessory minerals.

hackly — rough; jagged; broken.

holocrystalline — consisting entirely of crystals.

ilmenite — a usually massive, iron-black mineral of submetallic luster.

induration — hardening.

Kapton — insulating material used in the construction of the lunar module.

lithic — of, relating to, or made of stone.

mafic plutonic complexes — dark crystalline minerals characterized by magnesium and iron; formed by solidification of molten magma.

mascons — areas of mass concentrations of the Moon.

microlite — mineral consisting of an oxide of sodium, calcium, and tantalum with small amounts of fluorine and hydroxyl.

olivine — a mineral consisting of a silicate of magnesium and iron.

ophitic — lath-shaped plagioclase crystals enclosed in augite.

peridotites — any of a group of granitoid igneous rocks composed of olivine and usually other ferromagnesian minerals, but with little or no feldspar.

pigeonite — a monoclinic mineral consisting of pyroxene and rather low calcium, little or no aluminum or ferric iron, and less ferrous iron than magnesium.

plagioclase — a triclinic feldspar.

platy — consisting of plates or flaky layers.

pycnometer — a standard vessel for measuring and comparing the densities of liquids or solids.

pyroxene — a mineral occurring in monoclinic short, thick, prismatic crystals or in square cross section; often laminated, varying in color from white to dark green or black (rarely blue).

pyroxenite — an igneous rock that contains no olivine and is composed essentially of pyroxene.

regolith — the layer of fragmental debris that overlies consolidated bedrock.

shards — angular curved fragments.

subhedral — incompletely bounded by crystal planes.

terra — an area on the lunar surface that is relatively higher in elevation and lighter in color than the maria. The terra is characterized by a rough texture formed by intersecting or overlapping large craters.

traverse map — a map made from a series of movements that are joined end to end and are completely determined as to length and azimuth.

troilite — a mineral that is native ferrous sulfide.

twinned — refers to two craters with overlapping rims.

vesicle — a small cavity in a mineral or rock ordinarily produced by expansion of vapor in the molten mass.

vug — a small cavity in a rock.

APPENDIX B
Acronyms

ALSCC — Apollo Lunar Surface Closeup Camera
ALSEP — Apollo Lunar Surface Experiments Package
ALSRC — Apollo Lunar Sample Return Container
BRN — Brown & Root-Northrop
CAPCOM — capsule communicator
CM — command module
CRA — Crew Reception Area
CSM — command and service module
DPS — descent propulsion system
DTREM I — Dust, Thermal, and Radiation Engineering Measurements Package
EASEP — Early Apollo Scientific Experiments Package
e.d.t. — eastern daylight time
EVA — extravehicular activity
g.e.t. — ground elapsed time
JPL — Jet Propulsion Laboratory
KSC — Kennedy Space Center
LEC — Lunar Equipment Conveyor
LM — lunar module
LP — long period
LPZ — vertical-component seismometer
LRL — Lunar Receiving Laboratory
LRRR — Laser Ranging Retroreflector

MCC — Mission Control Center
MESA — Modularized Equipment Storage Assembly
MIT — Massachusetts Institute of Technology
MQF — Mobile Quarantine Facility
MSC — Manned Spacecraft Center
MSTPHC — multistop time-to-pulse height converter
NASA — National Aeronautics and Space Administration
ORNL — Oak Ridge National Laboratory
PET — Preliminary Examination Team
PLSS — portable life support system
PSEP — Passive Seismic Experiments Package
RCL — Radiation Counting Laboratory
SL — Sample Laboratory
SM — service module
SP — short period
SRC — sample return container
STP — standard temperature and pressure
SWC — Solar-Wind Composition
TDG — time-delay generator
TIM — time-interval meter
TV — television
USGS — U.S. Geological Survey
u.t. — universal time

www.ingramcontent.com/pod-product-compliance
Lightning Source LLC
Chambersburg PA
CBHW081723170526
45167CB00009B/3678